# DIESEL

## TECHNOLOGY
## AND SOCIETY
## IN
## INDUSTRIAL
## GERMANY

# DIESEL

## TECHNOLOGY AND SOCIETY IN INDUSTRIAL GERMANY

•

DONALD E. THOMAS, JR.

The University of Alabama Press

**Library of Congress Cataloging-in-Publication Data**

Thomas, Donald E., 1938–
  Diesel : technology and society during the German
industrial revolution.

  Bibliography: p.
  Includes index.
  1. Diesel, Rudolf, 1858–1913.  2. Mechanical
engineers—Germany—Biography.  3. Diesel motor—
History. I. Title.
TJ140.D5T48 1987      621.43′6′0924 [B]      85–24515
ISBN 0-8173-5170-1 (alk. paper)

To Patricia and Christopher

# Contents

# Illustrations

## Photographs

## Figures

# Preface

The beginnings of this study of Rudolf Diesel can be traced back to my doctoral dissertation on Houston Stewart Chamberlain. He and other *völkisch* writers in Germany often exhibited a profound distrust of modern science and technology, proclaimed them to be mechanistic and dead, and proposed to substitute in their place some form of living, organic "intuition" (*Anschauung*). This anti-scientific attitude led me to become interested in and to study German philosophies of technology. Thanks to a suggestion by Dr. Otto Mayr, now director of the Deutsches Museum, Munich, I began to examine the social philosophies of the inventor Rudolf Diesel and his son Eugen. The fruits of that research were published several years ago.[1] Rudolf Diesel's life had proved to be so interesting, however, that the logical next step was a full-scale investigation of his career. The results of that effort are presented in the following pages.

My efforts have profited from the help of many people and institutions. Anyone who has done original research on Diesel must acknowledge the generous assistance of the M.A.N. Werkarchiv (Augsburg-Nuremberg Engine Works Archive) and its staff. This collection contains the bulk of Rudolf Diesel's *Nachlass* (literary remains), as well as much supporting material. For their never-failing kind and courteous assistance and for their friendship, my thanks are extended to the present director, Herr Klaus Luther; to the former director, Frau Irmgard Denkinger; and to their staffs. A special debt is also owed to the staff at the Research Institute for the History of Science and Technology, the library, and the Special Collections Department of the Deutsches Museum, Munich, for providing access to the rest of Diesel's *Nachlass*. The museum's outstanding library on the history of technology provided many of the primary and secondary sources needed for this study. Special appreciation is accorded to Frau Margret Nida-Rümelin, Frau

Gertraud Ellerkamp, Dr. Michael Davidis, and the late Professor Friedrich Klemm of the Research Institute; Dr. Rudolf Heinrich of the Special Collections Department; Dr. Ernst Berninger of the library; and Herr Hans Strassl, head of the Street and Rail Transportation Division.

It has been a special privilege to have made the acquaintance and friendship of Herr Rainer Diesel, grandson of the inventor, and of his family. On numerous visits to Freiburg, he has opened his home to me as well as my family and made available the papers of his father, Eugen Diesel. My thanks are also tendered to Frau Dorette Breig, Stockholm-Solluntuna, Sweden, who shared her family recollections and let me use the memoirs of her mother, Hedy von Schmidt, Rudolf Diesel's daughter. The aid of Herr Diesel and Frau Breig have made it possible for me to gain a much deeper personal understanding of Rudolf Diesel than would otherwise have been possible.

Dr. Lynwood Bryant, professor emeritus at the Massachusetts Institute of Technology, has been extremely helpful with advice, the lending of supporting material, and his careful review of an earlier draft of the manuscript. Much of what I know about Rudolf Diesel and his engine has been provided by Dr. Bryant's insightful articles and comments.

Professor Lamar Cecil, of Washington and Lee University, in Lexington, Virginia, kindly read the manuscript and offered many helpful suggestions on content and style. I would also like to thank the readers for The University of Alabama Press, especially for their extensive comments on the Carnot cycle and engine efficiencies, which are discussed in chapter 3.

Versions of chapter 2 were read in 1983 at the Center for the Study of Science and Society at Virginia Polytechnic Institute and State University, in Blacksburg, Virginia , and in 1984 at the Research Institute of the Deutsches Museum. The discussions following each talk were very helpful to me.

For their assistance in various phases of consultation as well as the reading of grant proposals and parts of the manuscript, I am indebted to Professors Henry S. Bausum, Daniel C. Brittigan, D. Rae Carpenter, Willard Hays, and Edwin Dooley, of the Virginia Military Institute; Mr. Ganesh Bal, of the Pennsylvania State University; and Mrs. Carla Cecil, of the University of Virginia. Dr. Fred Hadsel, former director of the George C. Marshall Research Library, in Lexington, Virginia, and his staff made work space available to me that facilitated the writing of the manuscript. Head Librarian James E. Gaines, Jr., Mrs. Wylma P. Davis, and the staff of the VMI Preston Library were also most helpful, especially in obtaining interlibrary loan materials.

I have translated all German sources not otherwise available in English. Professor Klaus Phillips, of Hollins College, Virginia, was especially helpful in translating difficult German phrases, and Mrs. Trudy Fox, of Fairfield, Virginia, rendered invaluable assistance in translating letters written in the Gothic script.

Research and writing have been made possible by numerous grants: a Study Visit grant of the Federal Republic of Germany in the summer of 1977; a National Endowment for the Humanities (NEH) Summer Grant for Professor John Burke's 1978 seminar "Technology and Values in Twentieth Century America," at the University of California in Los Angeles (UCLA); a National Endowment for the Humanities Summer Stipend in 1980; and a National Science Foundation Summer Scholar's Grant in 1982. My special gratitude is expressed to the VMI Foundation and Research Committee, which awarded grants in 1977, 1980, 1982, and 1984. A VMI Wachtmeister sabbatical leave in the fall of 1982 allowed me to finish the first draft and write the second one.

Also appreciated are the efforts of Mrs. Conna Oram and Mrs. Lynn Williams, who proofread the manuscript; Mrs. Janet Cummings, who typed numerous drafts; Mrs. Janet Aldridge, who typed numerous grant applications; and Mrs. Monika Dickens, who typed appendix 2 in German.

My family and I tender our deep gratitude to Frau Elisabeth Schubert and Dr. Sylvia Schubert for their friendship and assistance on our many trips to Augsburg.

# DIESEL

## TECHNOLOGY
## AND SOCIETY
## IN
## INDUSTRIAL
## GERMANY

# INTRODUCTION

This book examines the life and work of Rudolf Diesel as a case study in the interrelationship of scientific, technological, social, and economic trends in late-nineteenth- and early-twentieth-century Germany.[1] During this period, Germany completed the transformation from a disunited, technologically backward nation to a united, industrialized, and technologically (if not politically) innovative one. Thus, the era is a vital segment of modern German history, and Diesel was one of the most important, if controversial, inventors during this period. He was twelve years old when the Franco-Prussian War broke out—an event that brought him from France to Germany—and he died less than a year before the beginning of World War I.

This study emphasizes three major areas. The first is biographical and is contained in chapter 1, though references to Diesel's life are made throughout the volume. The focus in that chapter is on the growth of the successful inventive personality, as well as the traits that led to Diesel's ultimate tragedy. Even those modern historians of technology who are engaged in building models of technological change indicate that in the end the individual plays a large role in this process.[2] No attempt, however, is made in this book to resurrect a romantic theory of the heroic inventor, nor is any implication made that Diesel either had no predecessors or that he was unrepresentative of larger trends. As much of an individualist as he was, he did work within well-established social and technological movements of his time.

The second major area of emphasis is the relationship of Diesel's social ideas to the larger topic of technology and culture in nineteenth-century Germany. This is addressed in chapter 2. Diesel's life spans a period when a new profession—engineering—was consolidating in his country. Engineers' desire for acceptance in their society met resistance from the older, established professions, such as law, and the humanistically oriented educational system. Hence, the acquisition of status became a leading concern of engineers. They were also discussing technology's importance for culture; some even advocated specific inventions to solve social problems. After an examination of the rise of the engineering profession and its interests, Diesel's ideas are discussed as both an example of engineering approaches to social problems and as an illustration of how invention can be motivated in part by social concerns.

The third major area of emphasis, discussed in chapters 3 through 5, is the process of invention, development, and innovation using the example of the diesel engine. The definition of these terms is taken from historians Thomas P. Hughes and Lynwood Bryant as well as the economist F. M. Scherer.[3] Invention is the intellectual process whereby the inventor conceives a new idea; development refers to the process of turning the idea into a working model; and innovation refers to the refinement of the invention and its introduction into a commercial market. Actually, all three of these processes may occur at the same time in a kind of dialectical interaction.[4]

Among the themes treated are the interrelationship of technology and science, the movement to replace the steam engine with a more thermally efficient engine, the impact of German patent law, the ways in which industrial backing was achieved, the relationship of inventor and entrepreneur, and the difficult road from development to innovation. The book ends with a Conclusion and a brief Epilogue, which surveys the history of the diesel engine since the first decade of the twentieth century. Although Diesel did not live to see it, his engine eventually became a major power source, especially in the field of transportation.

The elements of invention, development, and innovation constitute technological change. It is convenient to look at such change as a social process involving the flow of information and feedbacks among these three areas. The economist and historian Hugh Aitken has called the men and institutions who are the carriers of technological change "translators," who can transfer or translate knowledge from one area to another.[5] Diesel certainly fits the role of translator from scientific idea to invention and from invention to the working engine. However, he failed as a translator from development to innovation, that is, to a marketable product. Even his collaboration with Heinrich Buz, head of the Augsburg Engine Works, failed to produce a "team in which technological and commercial skills [were] synchronized."[6]

Diesel lived at a time of transition between the nineteenth-century inventor-entrepreneur, who was often only partly educated in science, and the twentieth-century scientifically educated inventor, who works in a governmental or industrial research lab. In Aitken's terms, the institutionalization of the translator role, the feedback among science, technology, and economics, was not as systematized during the earlier period as it was to be in the twentieth century. It is to be hoped that the Diesel story will not only help in understanding technological change in a past age, but will also contribute to knowledge about the foundations of our own technological era.

# CHAPTER 1

## Rudolf Diesel and the
## Nature of the Inventor

•

Rudolf Diesel is one of the most fascinating of modern inventors, a man whose complex and contradictory personality strongly influenced many of the early technical and business successes—as well as failures—of the diesel engine.[1] Although increasingly convinced of his destiny as a great inventor and able to sway others by the eloquence of his arguments, he was just as unable to see his own mistakes. An idealist in terms of his social thought, he could be quite deceptive in his business dealings or in public pronouncements about the status of his invention. Many of his friendships with those men who helped his cause ended in suspicion, distrust, and broken relations. He was a good family man, but he kept his wife and children in the dark about his growing financial difficulties and his ultimate decision to commit suicide. His tragic life is an interesting example both of the inventive personality and of the human psyche.

Rudolf Christian Karl Diesel was born on March 18, 1858, in Paris, of German parents. This "French connection" is important, for Diesel spent the first twelve years of his life in Paris and then returned to France between 1880 and 1889 in his first job after college. He spoke French so fluently that often he could not be distinguished from a native.[2] Even after his marriage in 1883 and return to Germany seven years later, French continued to be the language spoken at home by family members until the outbreak of World War I.[3] Diesel's experience in Paris helped to provide him a cosmopolitan view and perhaps also a taste for art and luxuries.[4] His manner was also affected; for example, he often used Latin rather than German spelling and, from the 1870s on, wrote in the Latin rather than the Germanic script that was still the standard in nineteenth-century Germany.[5] His knowledge of French utopian socialist writers and French intellectual movements of the end of the century undoubtedly played a major role in his social theories.

Diesel's ancestry can be traced back to the Thirty Years' War and consisted primarily of south German, Protestant, small craftsmen. In a time that featured little social mobility, his family was no exception. Unacceptable today is Eugen Diesel's assertion that his father was the product of years of "breeding" (*Züchtung*) of urban craftsmen and businessmen—one of the few concessions his biography makes to the prevailing zeitgeist of the 1930s in Germany.[6] Rudolf Diesel's grandfather settled in Augsburg; he and Rudolf's father were bookbinders. Theodor (1830–1901), Rudolf's father, emigrated to Paris around 1850, perhaps influenced in part by the unrest accompanying the revolutions of 1848. About the same time, several of his cousins moved to the United States. Apparently, Theodor also had visions of becoming something more than a craftsman like his father.

In 1855 Theodor married Elise Strobel, originally from Nuremberg, who had emigrated first to England and then, after returning for a time to Germany, to Paris, where she gave English and German lessons. Why Elise chose to leave Germany cannot be clearly determined. As Eugen Diesel indicates, increasing travel possibilities and wealth, as well as the desire to learn foreign languages, were generating a need for governesses and companions. Elise's letters to her parents describe the splendor and activity of London, but indicate she could never feel at home in that city. She and Theodor were married in London, apparently because papers were easier to obtain there.[7] Three children were born to the couple: Louise in 1856, Rudolf in 1858, and Emma in 1860.

Whatever Theodor's hopes, they remained unfulfilled. In Paris he had opened his own shop, which employed five or six workers and manufactured leather ware, portmanteaus, and letter cases. The business was never independent, because he always worked on consignment for larger export firms. Furthermore, the seasonal nature of demand resulted in several slack periods during the year, which made the family's income rather precarious. Elise was the practical member of the family, helping out in the shop. Theodor lacked business sense and was often caught up in impractical schemes.[8] After the Franco-Prussian War (1870–71), things went from bad to worse. In 1877 the couple sold their business and moved to Munich.

Theodor became more and more involved in Catholicism and spiritualism, especially after the death of his older daughter, Louise, in 1873.[9] He eventually decided that he possessed magnetic healing powers. From that time onward, he devoted himself to magnetic healing—a "medical" practice well established by that time. Under the title T. H. Diesel, "Practicing Magnetopath," he wrote a book, published in Munich in 1882: *Magnetotherapie: Der animalische oder Lebens-Magnetismus.*[10] In it, he explained that magnetism was a form of electricity circulating in the human nervous system. Those people who had an excess amount of magnetism could transmit it to those with a deficiency and thus heal illnesses, which were basically disturbances of the cell's electrical system. One of Rudolf's family scrapbooks contains a picture of Theodor "magnetizing" grandson Eugen. Unfortunately, this "delusional" quality, as Rudolf called it, was also an ingredient of his own character.[11] In any event, Theodor caused increasing problems for his son and may have helped hasten his abandonment of organized religion for the "religion" of science and technology.

Rudolf's early Parisian years were difficult and in later life he did not wish to speak of them.[12] Not only did he grow up in straitened circumstances, but also his father entertained peculiar Rousseauist ideas of education. These led him, for example, suddenly to trip his son or throw him into ditches without warning in order to teach him to cope with life. Punishment for misdeeds could be severe. For taking a clock apart, Rudolf was tied to a chair while his family went on a Sunday outing.[13] Later in life, he would apply some of this same strict discipline to his own children.

The sense of being a foreigner, a German in Paris, perhaps also alienated Diesel. Yet, his mother taught him English and he grew up in an urban, cosmopolitan atmosphere, aware of art and culture. Nevertheless, he apparently grew up essentially alone and embittered, with few friends. It was perhaps this background that caused him to generate compensatory mechanisms. At an early age he thought of himself as destined for great things and a high position that would distance him from his humble origins.[14]

In Paris, Diesel experienced his first contact with modern technology. As a youth of ten or twelve, he often visited the Conservatoire des Arts et Métiers, the oldest technical museum in the world. He saw, among other things, steam engines and the first self-propelled vehicle, N. J. Cugnot's three-wheeled steam wagon.[15]

In September 1870 the family was forced to leave Paris for London as part of a general expulsion order by the new republican government during the Franco-Prussian War. The Diesels were not allowed to return to Paris until the summer of 1871. In November 1870, after a brief sojourn in London, Rudolf was sent to Augsburg to live with relatives and to study at the commercial school there. His foster parents were Professor Christian Barnickel and his wife, Betty, a cousin of Theodor Diesel.[16] Rudolf remained in Augsburg until 1875, studying first at the royal commercial school and then at the industrial school. About this time, he finally decided on a career as an engineer and he so announced in a letter to his parents on his fourteenth birthday.[17] He rapidly acquired the traits of organizing every stage of his life and applying himself judiciously to whatever he wanted to do. His grades were the best in school. However, he also began to evidence a streak of intolerance for those he thought were either wrong or slower than he. Already, he was experiencing physical symptoms, such as headaches, that would increase in intensity in later years.[18]

Rudolf Diesel in 1870, age 12.
(M.A.N. Werkarchiv)

During his secondary schooling, Diesel became acquainted with a pneumatic lighting device, a so-called fire piston, which was in the possession of the industrial school. The fire piston was a glass tube containing a piece of tinder and a plunger. Compression of the air caused the tinder to glow. Whether or not this device served as an early stimulus to Diesel's later invention is uncertain. He referred to a similar device, the atmospheric piston, in his first hand-written patent draft to illustrate compression ignition.[19] Eugen Diesel reports that during the Second Power Machine Exhibition, in Munich in 1898, his father borrowed his old school's lighter, brought it to the exhibition, and used it to demonstrate to his children the principle of compression ignition.[20]

In the summer of 1875, Diesel took his final exams at the Augsburg industrial school and attained the highest marks of any pupil until that time. His outstanding academic work brought him to the attention of Professor Karl Max von Bauernfeind, who had been instrumental in the upgrading of the polytechnic school in Munich to a Technische Hochschule in 1868. On a trip to Augsburg, Bauernfeind personally examined Diesel and then offered him a two-year scholarship at Munich.[21]

Diesel's decision to become an engineer and receive higher technical education brought on conflict with his parents, who were weighed down by financial considerations and preferred that he finish school as quickly as possible so that he could work to support them. In one angry letter of 1874, Diesel criticized his mother in particular for being opposed to all science, especially mathematics. He also announced the probability of obtaining a scholarship.[22] Theodor's answer reproached his son for attacking his mother. "Didn't she try to awaken your interest with child's games and stories. Do you think a man is what he is simply because of himself or does the education provided by his parents play any role?"[23] Eventually, Diesel's parents had to accommodate themselves to his plans. By March 1875 Theodor was writing to Professor Barnickel in Augsburg, obviously pleased with the situation and indicating he had no objections to Rudolf's plans to attend the Munich Technische Hochschule.[24]

This was one of a number of engineering schools that had been established in Germany in the nineteenth century. Diesel entered the school's mechanical-technical division and studied there for four years. His program was technical, with two exceptions: German history and Goethe's poetry. During his last two years, he took several courses in theoretical machine design from Carl Linde and a laboratory course in the same field from Professor Moritz Schroeter.[25] Linde (1842–1934) had studied under such illustrious teachers as Gustav Zeuner, Rudolf Clausius, and Franz Reuleaux at the Zurich Polytechnicum. Linde was the founder of the modern refrigeration business and was the first to set up a practical laboratory for machine design in a German engineering college. He was ennobled for his achievements in 1897.[26] Linde was to be influential in Diesel's later career, inspiring him with the idea that led to the diesel engine, being his first employer, and presenting him with a model of inventor relationships with business that would be important in Diesel's attempt not only to gain industrial backing for but also to exploit his invention. Schroeter (1851–1925) would later become one of his principal supporters.

During his college years, Diesel's orthodox religious convictions began to weaken, his parents came to live in Munich, and he eventually tutored French to help support himself. He made several friends during these years, including Oskar von Miller (1855–1934), who was to be instrumental in the early development of electrical power in Germany and who remained Diesel's lifelong backer. Diesel also exhibited traits that would become pronounced in later years: a tendency to overwork and to become so absorbed in what he was doing that he could not unwind, which led to headaches and insomnia.[27]

Diesel had meant to take his final exams in July 1879, but was forced to postpone them because of a bout with typhus. After his recovery, he worked from October 1879 to the end of the year at the Sulzer Brothers factory, in Winterthur, Switzerland. He had been recommended to the company by his teachers, Linde and Schroeter. Sulzer built, among other things, Linde refrigerators, and Linde's idea was to provide some practical experience for his former pupil and then hire him for the new French Linde company.[28] Besides observing technical details in the factory, Diesel seems to have become interested in "the social question"—that is, the social problems caused by the industrial revolution and the working class's growing hostility toward capitalism. This interest would stay with him for the rest of his life.[29]

At the beginning of 1880, Diesel returned to Munich to complete his exams. He passed with the highest grades in the twelve-year history of the Munich Technische Hochschule and received a diploma in civil engineering.[30] In March he went to Paris to help set up the French affiliate of the Linde Refrigeration Company. He was at first hired as an apprentice for a modest salary. He worked for Baron Moritz von Hirsch, a well-known millionaire financier who was involved in the building of railroads in the Ottoman Empire and who provided much financial support to Eastern European Jews. Hirsch had purchased the rights to Linde's French patents and was constructing an ice-fabrication factory on the Quai de Grenelle in Paris near the site of the future Eiffel Tower. By 1881 Diesel had been promoted to factory director. After several years, Hirsch sold his patent rights back to Linde. Diesel then worked directly for Linde as his chief representative in France and as the director of the Paris refrigeration factory.[31]

These years were extremely busy ones for Diesel, as he traveled through France, Belgium, and even North Africa as a troubleshooter for Linde's company, repairing machines, taking part in meetings, and negotiating contracts. He also found time to embark on his own career as an inventor and apparently engaged in a number of love affairs, including one with an American divorcée artist, a Mrs. Fullerton, with whom he was supposedly very much in love and with whom he thought of returning to America. In a letter of December 5, 1880, to Diesel's sister Emma, who had apparently introduced the two, Mrs. Fullerton exclaimed: "He is a nice modest, refined young man! He has an artistic temperament—a poetic temperament—and a mind far above the generality of young men. I think I understand him pretty well."[32]

However strongly Diesel might have felt, this affair came to nothing. In late 1882 he met a young German girl, Martha Flasche, who was the daughter of a notary from Remscheid and governess to the children of a German merchant living in Paris. They were married in November 1883.

Diesel's letters to his fiancée in the several months before their marriage convey an interesting picture of the young engineer's sentiments and beliefs. By that time he had moved away from organized religion. He rejected all religious fanaticism as well as intolerance and championed the idea of helping people in this life rather than trying to convert them to belief in life in the next world. Part of this rejection was undoubtedly attributable to his dislike of his father's growing spiritualism. In June 1883 Rudolf wrote his fiancée, while she was visiting his parents in Munich:

> I know from experience that onesided spiritualist studies can turn men into real fools and, therefore, beg you to be cautious. . . . One does not have to be a spiritualist, a Protestant, a Catholic, or a Jew in order to feel true love of humanity in his breast and to practice it to the best of his ability. On the contrary, the less one adheres to specific opinions, the more one is open, the freer are his opinions and the more tolerant and loving toward his fellow men and the better is his heart. Our goal should not be to better the *future* happiness of men through intolerance and externals. Rather we should help our brothers on *this* earth, improve the situation of mankind, and redress poverty to the best of our ability. This seems to me better understood religion than that which neglects the earthly in favor of an unknown future. Jesus taught neither Protestantism, Catholicism, nor churchgoing and preaching. He taught love of mankind.[33]

Such religious views would stay with Diesel for the rest of his life, as he abandoned organized religion for a belief in science,

Martha Diesel in 1890.
(Special Collections Department, Deutsches Museum)

technology, reason, and progress. He would eventually propose the idea of a natural religion, in which the concept of God was nothing more than an expression of natural physical laws. Thus, he was well in tune with the rational progressive currents common among the late-nineteenth-century bourgeoisie. The spirit of this rational belief in progress was best reflected in the world expositions at the end of the century, of which Diesel saw a number, and in the museums of science and technology, typified by the Deutsches Museum, Munich, founded in 1903 by his friend Oskar von Miller.[34] This emphasis on helping one's fellow man is also related to the growing secularization of ethical ideals and to Diesel's own growing interest in the social question.

Furthermore, Diesel's letters to his fiancée show that he was one engineer who could be sensitive to the arts: "I am very happy that music makes such a deep impression on you. You know that I love music very much and that I consider people who have a deep understanding of music to be favored beings."[35]

After their marriage, the couple continued to live in Paris until the end of the decade. During these years, three children were born to the Diesels: Rudolf, Jr., in November 1884; a daughter, Hedy, in October 1885; and Eugen, in May 1889.

In the early 1880s, Diesel began his inventive career. His first patents, in the fall of 1881, were for machines that produced clear rather than crystallized ice. Although these machines were actually constructed, he eventually abandoned the project because by his contract he was bound not to compete with his employer or exploit refrigeration machines independently of him:

> My refrigeration machine gives me no hope for the future. Since my life at present is dependent on Linde, I can in no way appear as a rival. . . . So I have thrown the machine and all my private studies on the subject completely overboard, and I have turned in other directions.[36]

In late 1883 Diesel left his position as an employee of Linde. However, as an "independent merchant" and "civil engineer," he continued to represent Linde's interests in France and to head his

Paris factory. This made him somewhat more independent, but still not totally free of his obligations to Linde.[37]

From as early as 1880, Diesel had conceived the idea of finding "a more economical engine than our present power machines."[38] From 1883 to 1889, he spent much time working on an ammonia vapor engine, even constructing working models. He thought of it as a small power source that would compete with the steam engine. The engine was ultimately a failure, but it was the direct predecessor of the diesel engine.[39]

Also existing from this period is a brief sketch for a solar-powered motor. Air inside an iron container would be heated by the sun and, expanding, would drive a piston. Then the air would be cooled by injecting water. It would contract and atmospheric pressure would drive the piston back. Diesel quickly calculated that such an engine would not generate much power, perhaps 1/50 horsepower (hp), and was therefore "too weak for any kind of efficiency." The idea remained a four-page sketch, but it shows Diesel's mind was following a number of tracks at this time, all aimed at increasing the efficiency of engines.[40]

Despite these setbacks, Diesel continued to believe in his mission and began to complain about the restraints placed on him by his job. In a letter of July 28, 1884, he exclaimed to his wife: "I've had it with repairing machines. I'm short of temper even when I shouldn't be." On December 25, 1887, he informed her: "I don't have anything special in mind, but I definitely feel that my connection with Linde's company is rapidly coming to an end and that I must undertake something new in the future—but what?" On March 18, 1888, his thirtieth birthday, he wrote his parents:

> I'm now thirty years old. I appear to myself to be an old man and am basically sad that I have worked at and accomplished so little, for when I consider what I have done so far, it doesn't amount to much. On the other hand, I feel that my talents are really developing now, so that I haven't given up all hope.[41]

Diesel's hectic schedule and his own inventive efforts combined to cause a strain on his health, crystallizing character traits that had been evolving since his school years. Complaints about headaches, insomnia, and compulsive intensity in what he was doing crop up repeatedly in his letters from the 1880s. For example, in one to Martha of February 26, 1886, he commented, "Again I am having terrible headaches—I believe it comes from the food. Everything is fatty and strongly seasoned." On November 27, 1886, he again complained, "I am tormented by severe insomnia that makes me feel like I'm living in a dream." On July 20, 1887, he wrote to her again, concerning his work on the ammonia engine:

> I'm gone the whole day, eat lunch out, and return late in the evening. After dinner I write down my impressions of the day. I'm not sleeping at all well; I lie awake half the night and mull things over and over in my mind. If it is only a success, it will seem like salvation to me after years of imprisonment. I'm now living in desperate agitation.[42]

Diesel's propensity for overwork and associated symptoms, such as headaches and insomnia, would increase during his lifetime. Sustained periods of effort, like the time of the development of the first working diesel engine (1892–97), would be followed by nervous and physical collapse, necessitating recovery in a sanitarium in late 1898 and early 1899. A second sanitarium stay was necessary in 1901–2.[43] During his work periods, Diesel was often extremely optimistic, expressing exaggerated belief that his ideas were correct or that his cause would succeed. During his later years, when his creative, inventive period had passed and his financial affairs were going from bad to worse, he was able to convince himself that success was just around the corner. When the truth of his impending financial collapse could no longer be avoided, he took his own life. During his creative periods, he was so sure of himself and his mission that he was able to carry other people along with him. His father's growing delusions at the end of his life have already been pointed out.

Historical figures cannot be psychoanalyzed with very much certainty, but Diesel's symptoms indicate that he had a manic-depressive personality. Manic-depression is considered to be hereditary and is characterized by periods of intense concentration and activity, often accompanied by headaches, insomnia, and delusional feelings of success, alternating with periods of depression and possibly collapse. Although such a conclusion about Diesel may explain a number of his successes and failures, it is not meant to be an example of psychological reductionism. Whatever the psychological origins of some of his ideas and actions, they must be judged ultimately on their own merit.[44]

Added to Diesel's dissatisfaction with his job and his health problems were the growing difficulties that the political climate in France in the late 1880s, especially the Franco-German antagonism and the Boulangist movement, posed for his business. In a letter to his parents of January 28, 1887, he described the rumors in Paris of a new Franco-German war and criticized Bismarck for his part in keeping the war scare alive. He then went on: "I'm really not that worried, however, for the engineering profession has the advantage of enjoying a certain demand under all conditions . . . I must say that things here seem so unstable that I should always like to be ready to emigrate."[45] Diesel viewed the July 14, 1888, Bastille Day parade and in a letter to his wife described the Boulangist demonstrators in the crowd who booed Marie François Carnot, president of the republic, and other governmental officials.[46]

In 1889 Paris was the site of a Universal Exposition, and Diesel was involved with the exhibition of Linde machinery. In February, Diesel confided to his father: ". . . perhaps, if it [the ammonia engine] is completed—my own invention [will be exhibited], the child on which I have labored for five years and which now finally will be born. This is, however, a state secret, and I beg you not to discuss it with anyone."[47] In April he wrote to his parents, describing the construction of the Eiffel Tower and mentioning that he often saw Gustave Eiffel, who was head of the local engineering society in Paris.[48]

By April, however, Diesel had apparently suspended tests on the ammonia engine. Too many problems were still being encountered with efficient operation,[49] and this was likely the reason he decided not to try to show it at the exposition. About this time—late 1889 and 1890—the decisive shift occurred in his mind away from the ammonia engine and toward the idea of what would become the diesel engine.

The decade of the 1880s had been extremely important in establishing the path Diesel's future career would take. He had married and established a family, his inventive career had begun, and his interest in the social question had grown. Indeed, his main interests were, first, constructing a small engine that would compete with the steam engine and revitalize the artisan class; and, second, answering the social question.[50] If his efforts with the ammonia engine had been unsuccessful, his research and experience were leading him in the direction of the diesel engine. Despite setbacks, many of his letters from these years radiate the confidence of a man destined for greatness.[51]

In late 1889 Diesel decided to leave France and become Linde's representative in Berlin, though Linde did not want Diesel dividing his time between company matters and attempts at invention. Diesel thought that there he would be free of the political tensions and concern about possible expulsion that existed in France. Also, his chances to make connections with German industrialists who might back his inventions would be enhanced. Writing to his wife after their move, he said that his positive expectations about the city were fulfilled and that he had already met several businessmen and directors of engineering firms—in contrast to Paris where he was totally on his own. "The seed of my ammonia engine is beginning to germinate [into a new engine], perhaps the plant will grow faster than I think."[52]

The period from 1890 to 1897 was the high point of Diesel's life and work. Many of the events of this period relating to the invention and development of the diesel engine will be discussed in later chapters. Here it is sufficient to note that during the period

1890–92, Diesel continued to work out the theoretical basis of what was to be the diesel engine. These principles were included in a manuscript, finished in early 1892. Most of it, along with later additions, was printed in 1893 in Diesel's first book, *Theorie und Konstruktion eines rationellen Wärmemotors* (*Theory and Construction of a Rational Heat Engine*).[53]

During 1892 and 1893 Diesel obtained the backing of Heinrich Buz and his Augsburg Engine Works and then the Krupp Works for his invention, a high-compression internal combustion engine that would have a high thermal efficiency, that is, one in which almost 73 percent of the heat energy of the fuel would be converted to useful work. He obtained a patent for a "working process and its method of execution for internal combustion engines." Then, at the Augsburg Engine Works, he began the first series of attempts to construct a working engine. Although originally optimistic that the development phase would be relatively short, he found himself involved in an exhaustive series of time-consuming and tedious tests, full of false starts. Not until 1897 was a satisfactory working model of the diesel engine developed, tested, and pronounced a marketable product. However, the 1897 engine, though the most thermally efficient internal combustion engine of its time, deviated markedly from Diesel's original ideas. He nonetheless reported on his success in a speech to the German Engineers' Association at its meeting in Kassel on June 16, 1897. He was followed by his former teacher, Professor Moritz Schroeter, who had conducted the official tests on the engine in February and who now proclaimed it a "fully marketable product."[54]

In 1893, after Krupp had agreed to pay him 30,000 marks yearly salary during the testing period, Diesel had left Linde's company. He and his family continued, however, to live in Berlin; this involved the inconvenience of traveling back and forth from Berlin to Augsburg, which necessitated long absences from his family. In 1895 they moved, not to Augsburg, but to nearby Munich. At various times, Diesel argued that adequate housing could not be found in Augsburg, or that his situation might be seen as an anomalous

Rudolf Diesel in 1894.
(M.A.N. Werkarchiv)

one because he was not an actual employee of the Augsburg Engine Works. In a letter to Martha, however, he indicated his real feelings:

> In Munich one has the beautiful, easily accessible surrounding environment, a cosmopolitan atmosphere, museums, wonderful art exhibitions, theater, the intellectual excitement of a large city in which art and science flourish. One has the possibility of intercourse with intellectually significant people. In Augsburg—nothing, nothing, nothing.[55]

In December 1895 the relocation to the Giselastrasse in the artistic quarter of Schwabing was accomplished. In 1897 the family moved to the Schachstrasse near the Siegestor.

The years 1897 and 1898 mark the turning point in Diesel's career. Outwardly a successful inventor, he became the center of an international patent and licensing network to exploit his engine commercially. In the fall of 1897, for example, he signed a contract with the wealthy German-American brewer Adolphus Busch for the rights to his American and Canadian patents. Busch paid 1,000,000 marks, or about $238,000. In the summer of 1898, the diesel engine pavilion formed the focal point of the Second Power Machine Exhibition, in Munich, and in 1900 the diesel engine won the grand prix at the World's Fair, in Paris.

To reflect his new status as a successful inventor who was rapidly becoming a millionaire, Diesel began the building of a villa that was to be one of the most elegant residences in Munich. The architect was Professor Max Littmann. Diesel involved himself in all aspects of design as well as furnishing and demanded the best of everything. After two years of construction, the villa at Maria-Theresia-Strasse 32 was occupied in 1901. Land, construction costs, and furnishings added up to some 900,000 marks.[56] The villa was one of the best furnished homes in Munich, but it had cost so much it helped to undermine Diesel's finances.

The villa was eclectic in style. A huge entrance hall, designed partly in the German Renaissance manner, rose two stories and contained a broad staircase featuring a solid oak balustrade. The ground floor included a garden room with a fountain; a rarely used Louis XV salon; an elegantly furnished living room that was Martha Diesel's study and the center of family life; and Diesel's study, done in the then-popular *Jugendstil*. Upstairs, besides bedrooms, were a billiard and trophy room as well as a gallery around the entire hallway. The corridors and rooms were full of Renaissance cupboards, rococo desks, paintings, and a variety of other furnishings. The house contained five bathrooms and a huge cellar, in which the children rode their bicycles. The house was surrounded by a garden and situated opposite a park that was frequented by a number of Munich's famous residents. These included the Wagnerian director Felix Mottl; the actor Ernst von Possart; and the aged Helene von Dönniges, over whom the founder of German socialism, Ferdinand Lassalle, had lost his life in an 1864 duel. Occasionally, the elderly Bavarian Prince Regent Luitpold would have his carriage stopped at the corner and take a short walk past the Diesel house.[57]

The villa was the scene of many receptions and gatherings. However, for one family member, Eugen Diesel, it took on sinister connotations, coming to represent his father's declining fortunes and the gradual estrangement of his parents—a place he left forever in October 1913 without regret. His feelings may have been influenced by his father's tragic end. His sister, Hedy, was not at all negative about the Munich house, and in her memoirs she looked back with nostalgia on her youth there.

Diesel's fortunes were indeed on the decline. Already his invention was becoming the center of controversy and legal battles to try to invalidate his patent. From now until his death, he would have to put up with critics who tried to prove that the working diesel engine had little to do with his original ideas, or that he was more of a businessman than a creative inventor. Further, during the strenuous inventive period, 1893–97, he had seriously strained his health, aggravating his tendency to overwork, insomnia, and headaches. By 1898 he was on the verge of nervous collapse and

Diesel's *Jugendstil* workroom in his Munich villa.
(M.A.N. Werkarchiv)

that fall entered a sanitarium to recover his health. Not until April 1899 was he able to take up work again, missing a critical period in the innovative phase of the diesel engine.

During the period 1898–99, it became clear that the engine was not in fact a marketable product but was still beset with numerous "childhood ailments."[58] Engines were returned and licenses given up. The engine's reputation sank to an all-time low.

During the summer and fall of 1898, sensing his oncoming health crisis and fearing he might be permanently incapacitated or even die, Diesel had taken the lead in forming the General Society for Diesel Engines (Allgemeine Gesellschaft für Dieselmotoren), an organization in which Augsburg and Krupp participated and which essentially took over direct control and further exploitation of Diesel's patents and invention. He was on its board of directors until 1906. A large cash and stock settlement supposedly secured his financial future.

Although later during the first decade of the twentieth century, Diesel would attempt to renew his inventive successes—for example, by setting up his own design bureau in his house and collaborating with Sulzer Brothers, in Switzerland, on diesel locomotive engines—he never repeated his earlier triumphs. Basically, further development of the diesel engine was being carried out by such factories as the Augsburg-Nuremberg Engine Works (M.A.N.), in Augsburg, and its engineers, like Immanuel Lauster.[59] Such effort overcame the engine's early problems and turned it into a marketable product. By 1906 Diesel had even fallen out with M.A.N. and the General Society, which led to lawsuits against him. The irony of his situation during the last decade of his life was that the more he and his invention were honored, the less this fame was accompanied by financial gain.

Diesel's early licensing successes had made him a millionaire several times over. Unfortunately, much of his wealth was paper, such as stock of the General Society, which failed to show any profit. His house had been expensive. Further, he involved himself in unwise speculations in Galician oil fields and in the purchase of real estate in and around Munich, both of which ventures cost him heavily. Because an expensive lawsuit against the company from which he had purchased the real estate was lost, he was burdened with the payment for the land as well as the real estate taxes. Even exchanging some of the land for rental properties made no difference in his increasingly difficult financial affairs.[60]

Diesel was not an astute businessman. He did not know how to recognize and minimize losses. Eugen Diesel states that he was "penny wise but pound foolish."[61] He possessed a sense of moral righteousness and stubbornness, which caused him to plunge even further into a bad business venture because he was convinced he would be able to turn a loss into a profit. The end result was even further loss. He would not consult bankers or other advisers regarding his situation, nor would he approach those of his friends, such as Adolphus Busch or Oskar von Miller, who might have been able to rescue him financially. Late in life, he was putting money into a speculative enterprise in Munich to develop electric automobiles and into a Catholic lottery society, both of which lost money. By the summer of 1913, he finally realized he was on the verge of bankruptcy.[62]

Outwardly during those years—the first decade of the twentieth century—all was calm and orderly in the Diesel household, which displayed all the characteristics of the solid haute bourgeoisie. Diesel's mother had died in 1897. His father passed away in 1901, freeing Diesel from his long concern over Theodor's involvement in spiritualism and magnetism. Diesel's unwillingness to tell his wife Martha or his family of his growing financial difficulties, and his feeling until the very end of his life that he could recoup his losses and that success was again just around the corner, combined with his incredible self-control to create an atmosphere of business as usual.

Eugen Diesel's biography gives some generalized hints of the elegant and cultivated life in the Diesel household in Munich, but this life is portrayed more strongly and nostalgically by Rudolf's daughter, Hedy, in her unpublished memoirs. The children had governesses and received dancing and art lessons. The home was the scene of private Fasching balls. In addition, the family attended numerous parties and festive celebrations, organized by such artists as Franz von Lenbach and Franz von Stuck. Season tickets facilitated attendance at the opera, concerts, and the theater; daughter Hedy saw Isadora Duncan make her Munich debut. The Diesels were friendly with the Oskar von Miller and Carl Linde families and received many foreign visitors, especially those connected with the diesel engine business, such as Adolphus Busch and Jakob Sulzer-Imhoof. In 1901-2 Hedy was sent to a finishing school in Switzerland. In 1905 she was engaged briefly to the cousin of the later Nazi leader Hermann Goering.

The family also traveled extensively, especially to Austria, Italy, Switzerland, France, and England—always staying in first-class accommodations. Diesel purchased his first automobile in 1905, a seven-seater N.A.G.,[63] and in the summers the family, always driven by a chauffeur, took numerous trips, which were rather adventurous, in view of the state of roads and lack of service facilities at the time. Eugen Diesel remarks that, though his father justified the purchase by saying he needed to become familiar with all facets of an automobile before he could work on an engine for it, he was neither much of a mechanic nor a very good driver.[64]

According to Hedy, family discipline was strict, reflecting Diesel's own difficult upbringing, though never as harsh as what he had experienced. The brunt of his expectations fell on his oldest son, Rudolf, Jr., who, by Hedy's account, possessed a sensitive, artistic, melancholic temperament. He apparently never fulfilled his father's expectations, became involved in a bad first marriage, and died near the end of World War II.

In 1908 Hedy met a young engineer, Arnold von Schmidt, the son of a professor at the Munich Technische Hochschule. His grandfather, Friedrich Wilhelm, had been ennobled by Emperor Franz Joseph in 1886 for numerous accomplishments, such as the rebuilding of the tower of St. Stephens Cathedral in Vienna and the renovation of the Sistine Chapel in Rome.[65] In June 1909 Arnold and Hedy were married in an elegant wedding ceremony. The Diesels must have been happy to see this union of the new nobility with the technological elite. The couple honeymooned in Baden-Baden, complete with a N.A.G. auto and chauffeur provided by Diesel. On one outing, the car took a hairpin curve too rapidly and ran into a rock. Although both Hedy and the chauffeur were thrown from the auto, fortunately neither of them was seriously injured. In October, Diesel invited Arnold to join his design bureau in Munich. The young man accepted, but was never happy and in 1913 took a position with the Adler Works, in Frankfurt am Main.[66]

Diesel's expectations then fell all the more on his youngest son, Eugen, whom he hoped would become an engineer and follow in his footsteps. More and more influenced, however, by the cultural attractions of Munich and by his father's interest in social prob-

Eugen Diesel (1889–1970), biographer of his father and author of numerous works on technology and culture. (Special Collections Department, Deutsches Museum)

lems, Eugen gradually rebelled against the idea of an engineering career and by 1913 had switched studies to geology. Family letters from this period reveal that Eugen feared he could never be as successful an engineer as his father. Evincing youthful idealism, he also rejected the idea of continuing his father's work of "commercial and financial exploitation."[67] After World War I, Eugen became a professional writer, dealing with political and social issues as well as the impact of technology on society and eventually publishing a biography of his father in 1937. During the 1920s his political views were close to those of the anti-Weimar Republic conservative revolutionaries, but during the Nazi period he became a member of the resistance movement.[68]

At least as far back as his apprentice days in the Sulzer Brothers factory, Diesel had begun to manifest interest in the social question: the social ills of the industrial revolution, the condition of the working class, and the growth of class conflict in late-nineteenth-century Europe. Around the turn of the century, he spent some time collecting notes and statistics for a book on the subject, published in 1903 as *Solidarismus: Natürliche wirtschaftliche Erlösung des Menschen* (*Solidarism: The Natural Economic Salvation of Mankind*).[69] The book preached the solidarity of the individual and the community. It advocated worker-run factories that would provide a complete scheme of social welfare measures. Diesel thought he had answered the social question rationally, just as he had provided a rational solution to the problem of the internal combustion engine. Despite his high hopes, the book and his other efforts in this area made no impact.

Twice during the first decade of the twentieth century, in 1904 and again in 1912, Diesel journeyed to America. In 1904 he and his wife attended the St. Louis World's Fair and observed the status of the diesel engine business, especially with Adolphus Busch's company. Diesel's travels took him from Boston through the Midwest to San Francisco. He kept an extensive diary of this trip, in which he said that he was giving personal impressions of things that interested him, especially technical objects.[70] The diary is organized around subject headings, such as trains, cities, and women, rather than by date. True to his methodical nature, Diesel kept rather meticulous notes; for example, he indicated the weight, number of tons of coal used as fuel, and speed of his ship on the return trip.

The diary in the main presents a highly critical view of America. Life there, said Diesel, revolved around the dollar—anything that would save labor or time, that was expedient, and that would further the end result. Things were not done "if there is no dollar in it." The individual's freedom to do what he liked was nothing but an excuse for some to exploit and to repress others. The result of this philosophy was that all interest centered on technology, not on the arts or on an object's aesthetic qualities. Americans lived on canned goods, which were not only poor food but which were also an example of the way business exploited the housewife. Child labor, though officially outlawed, still existed covertly. American women were fat and lazy. The cities all looked alike. The business districts were full of noise and traffic, the quarters where blacks and Chinese lived looked dirty and poverty-stricken, and the many saloons were filled with a wild-looking clientele.

People were becoming more and more bitter about the Rockefellers, Carnegies, and their money. Most of the populace were raw, ruthless, and poorly educated in public affairs. Men often spit in public and put their feet up on the furniture, even when women were present. Smoking and chewing tobacco were popular. Elbowing and jostling from all quarters were to be expected in the streets. In the Far West, people were really "wild," carrying guns in the open. Safety standards did not seem well maintained, as evidenced by the numerous railroad accidents. Service in hotels and in restaurants was poor because of the black help. The St. Louis Fair was in general poor, presenting mostly objects of local curiosity and few foreign exhibits—a downward trend in world exhibitions that Diesel felt set in after the Paris Exposition in 1889. The Boston Art Museum was not very good and contained few German works, though private collections in the United States were said to be adequate.

Diesel was not, however, entirely negative. He praised the public and university libraries in the East, especially at Harvard and Columbia. He thought highly of Columbia's practical laboratory for testing locomotives. Despite his many criticisms, he contended that America's economic strength lay in the high wages paid to its workers.[71]

European visitors often attack America's lack of culture and complain about her materialism. It may seem odd, however, that Diesel, who believed so much in progress through technology, should have been so negative about American society. Perhaps Eugen is correct when he says that his father was too European, reserved, and earnest to respond to the openness in the United States and that he saw the country as a kind of deviation from the norm of technological progress.[72] Was this trip, as Eugen suggests, the beginning of Diesel's skeptical attitude toward technological solutions—an attitude that really only crystallized at the end of his life? Given his grand airs, his cosmopolitan background, and interest in the arts, it is not surprising that he reacted negatively to much of what he saw in American society. Perhaps it represented to him runaway technology without culture.

Eight years later, in 1912, Diesel returned to America from March to May. Although not an official member, he traveled with a study commission of the newly founded Deutsches Museum, in Munich, which was gathering information on American technical libraries and laboratories as well as trying to interest American industry in supporting the museum. One of the commission's members was Diesel's longtime friend and founder of the museum, Oskar von Miller. Diesel's real object in traveling to the United States was to conduct a promotional tour for the diesel engine. This time, his travels were limited to the cities of the East, such as New York, Philadelphia, and Boston, as well as a stop in Chicago. He was by now well known, and his engine, if not his personal fortunes, was out of its growing pains stage and winning more and more successes. He received much attention from the American press, which printed numerous articles on him and announced the arrival of the new age of oil.[73] Indeed, the trip took on aspects of a triumphant procession. In April news reached America of the sinking of the *Titanic*. According to Eugen Diesel, his father had at one time contemplated making the trip to America on that ill-fated liner.[74]

On May 6, the day before his departure for Europe, Diesel met with Thomas Edison at Edison's home in West Orange, New Jersey. There exists a photograph of the two inventors sitting in front of Edison's writing desk, which was crammed with papers. Diesel apparently found this a congenial visit. According to Eugen, Edison discussed with Diesel his various methods of invention and his own European trip, in which he found Italy boring because of all the

Rudolf Diesel and Thomas Edison
in West Orange, New Jersey, 1912.
(Special Collections Department, Deutsches Museum)

churches.[75] Edison played his latest phonograph for Diesel. His parting words were: "Don't eat too much."[76] On the following day, Diesel and his wife set sail for Europe. He noted in his diary: "Terrible rain, fog, a sad departure."

In November 1912 Diesel delivered a lecture on the development of the diesel engine before the German Society of Naval Architects, in Charlottenberg. During the discussion that followed, he was attacked vigorously by such men as Alois Riedler, professor at the Technische Hochschule Berlin-Charlottenberg and prolific writer on curriculum reform as well as the relationship of technology and culture. Riedler argued that Diesel's theory had never been realized in the working engine and that the actual diesel engine was much more the product of M.A.N. engineers. The audience was clearly on the side of Diesel during the debate.[77]

As the year 1913 progressed, Diesel's financial situation continued to deteriorate rapidly. He finally recognized that a financial collapse could not be prevented and would come sometime during the fall of that year. He was still too proud to ask for money or advice from friends or to let his family know of the situation. On several occasions, he suggested either that they sell their house or emigrate to the United States, where Diesel could probably obtain a position as a consulting engineer. However, because he did not explain to his wife the true reasons for such action, she rejected both ideas.

During this period, critical works examining the development of the diesel engine and Diesel's role as the inventor continued to appear. One of the most violently critical, indeed one that is dismissed today as a distorted personal attack, was that of Professor Johannes Lüders, of the Aachen Technische Hochschule. Diesel had learned from his own publisher, Julius Springer, what the nature of the book would be. Among other things, Lüders accused Diesel of outright lying, lacking sufficient technical background for his inventive efforts, and contributing little to the developmental period, 1893 to 1897.[78]

No doubt these continuing attacks on his reputation added to his depression, though they alone were not responsible for his tragic end. No wonder he would say of the innovation period in his *Die Entstehung des Dieselmotors* (*The Origins of the Diesel Engine*): "[It is] a time of struggle with stupidity and envy, disguised resistance and open struggle with interest groups, a terrible time of struggle with other men, a martyrdom, even when one is successful."[79] Diesel was apparently also depressed by his son Eugen's final break with engineering studies and his decision to study geology, though he arranged a financial stipend out of his last remaining funds to help his son conclude his studies.[80]

During the summer of 1913, Diesel continued to work as usual, even holding a reception for more than a hundred visiting American engineers in June and attending an international building exhibition in Leipzig, where he took a trip on the zeppelin airship *Saxony*.[81] Also during that summer, he accepted an invitation to travel to England in late September to attend the ground-breaking ceremonies for the British Diesel Company's new plant at Ipswich and to address the Royal Automobile Club in London. He carefully attended to all correspondence and plans for the visit. His two sons were temporarily absent from Munich, and in the middle of September he sent his wife off on a trip to her relatives. He was now alone in his villa, which he had come to think of as a "mausoleum," housing himself, "a living corpse."[82] On Sunday, September 15, he burned a number of papers, primarily related to his financial and business dealings. The major documents relating to his engine had already been given to the Deutsches Museum. Obviously, he was preparing to end his life, though exactly when the notion of suicide became concrete is impossible to say.[83]

On September 20 he traveled to spend a few days with his son Eugen, who was carrying out geological studies near Nördlingen. They then journeyed to Frankfurt, where Diesel spent several days with the entire family, staying with his daughter, Hedy, and her husband. During this time, while visiting the Adler Works factory, where his son-in-law worked, Diesel made his oft-quoted statement: "It is beautiful for an engineer to shape and design the same

Rudolf Diesel near the end of his life.
(Special Collections Department, Deutsches Museum)

way that an artist shapes and designs. But whether the whole process makes any sense, whether men become happier—that I can no longer decide."[84]

On September 26 Eugen saw his father off to Belgium, where he joined George Carels, chief of the Carels Brothers factory in Brussels, and Carels's chief engineer. On the 29th of September, the three men boarded the steamer *Dresden* in Antwerp, for the crossing to Harwich, England. Diesel's last letters and cards to his family carefully described his impressions of Belgium and his exact itinerary. At the same time, he mentioned that he was reading Schopenhauer and also complained of bad health. For example, in a letter to Rudolf, Jr., he said: "For a while things have not been going according to wish. My heart is giving me difficulties. Many times it seems to me almost to be stopped. I also have remarkable shooting pains in the head and my general disposition is poor."[85]

A similar letter had been sent to his wife, Martha, on September 27, but it was misaddressed and arrived after his death. Eugen argues that his father may still have had ambivalent thoughts about the decision to end his life. On the one hand, he may have been trying to alert his wife so that she would come to his aid. On the other hand, by misaddressing the envelope, he may have sought to ensure that she could not interfere.[86]

About 10:00 P.M. the travelers retired to their cabins. The following morning, September 30, 1913, it was discovered that Diesel had disappeared from the ship. Testimony given later by Carels indicated that Diesel had seemed to be in good spirits on the 29th and spoke enthusiastically about the future of his engine. All personal effects in his cabin were left in perfect order. In early October, a Dutch pilot boat recovered a corpse from the sea; removed certain personal effects; and, as was the custom, committed the body back to the sea. Inspection of these effects, verified by Eugen, revealed that they had belonged to Rudolf Diesel.

The exact nature of Diesel's death will probably never be known, but Eugen is of the opinion, and all facts seem to confirm it, that he took his own life. His impending bankruptcy, combined with his poor health and the continuing attacks on him, certainly provided reason enough.[87] To the end, he was too proud or ashamed to tell his family and friends the true nature of his situation. Ironically, he might have been able to save himself financially. Adolphus Busch, himself seriously ill at the time, is reported to have said on hearing of the death: "If only my friend Diesel had said something to me. I knew that things were not going well with him."[88] To the end, all details of his personal life and correspondence were carefully worked out. It is as if he had put on an act for so long that he compulsively carried out his part to the end. Various sensational accounts soon appeared; for example, one alleged that Diesel had been murdered by German agents so that he could not impart technical secrets on diesel submarine engines to the British. None of these accounts has ever been substantiated.[89]

In summary, a number of personality traits help explain Diesel's successes as an inventor and, alternatively, his failure in later life as a businessman. Joseph Rossman has isolated a number of characteristics typical of the successful inventor, all of which apply to Diesel. They include intelligence, tenacity, imagination, suspicion, lack of business ability, and belief in the individual's mission and success.[90] Most important in relation to Diesel is what Rossman calls "invention as self-expression."[91] "Invention, according to this view, is caused by the desire to compensate for one's inferiority." In Diesel's case, inferiority feelings were not caused by a physical handicap, but by a childhood involving difficult parents and straitened circumstances. It has been suggested that Diesel reacted against his early life by forming conceptions of his calling to greatness and the desire to excel in his studies and profession. It has also been suggested that he had a manic-depressive personality. The combination of intelligence, fierce work habits, motivation to succeed, and verbal persuasiveness—qualities no doubt in part heightened at times by his psychological state—all contributed to his inventive achievements. On the other hand, his fixation on success, inability to relax, insomnia, propensity to headaches, and paranoia concerning his colleagues not only led eventually to physical and mental collapse, but also opened the way to business failure and suicide.

Does Diesel fit the pattern of the malcontent, who invents to shake up a society he does not like? This is what historian Elting Morison calls "invention as a hostile act."[92] Certainly, enough evidence exists to suggest that Diesel wanted to be accepted into society to gain social status, even to gain a considerable amount of wealth from his invention. He once wrote to his wife that, unlike most millionaires, he slept better the richer he became.[93] On the other hand, he disliked many facets of the society of his age. Further, he was sympathetic toward the working class. As he wrote in 1910 to his son Eugen, then working as an apprentice:

> . . . you will learn more there about the truth of life and social relations than in our circles which have absolutely no idea of such things and regard every worker as a kind of robber. . . . I regard it as good fortune, when a young man has the opportunity to get a glimpse into all social classes and recognizes that the lower rungs are better than the higher.[94]

Diesel's attitude toward society, then, was ambivalent. An understanding of this ambivalence and the role social factors played in the development of the diesel engine requires an analysis of his social ideas within the context of the relationship of technology to culture in nineteenth-century German society.

# CHAPTER
## 2

# Diesel, German Engineers,
# and the Social Question

•

Rudolf Diesel's involvement with the social question must be seen against the background of larger economic, social, and technological trends in late-nineteenth-century Germany. Many of the concerns expressed by engineers there—such as those related to status or growing class division—were also mirrored in other national engineering traditions, for example, the American.[1] This chapter will examine the German background of the problem and then look at Diesel's ideas as a case study in the social thought of that country's engineers.

## Engineers and German Society
## in the Nineteenth Century

From around the middle of the nineteenth century on, the process of the industrial revolution began to accelerate rapidly in Germany, helped especially by political unification in 1870–71. This second phase of German industrialization was based on such industries as coal, iron, steel, chemicals, and machinery. Whereas during the first half of the nineteenth century England had predominated industrially and Germany had imported technology, mostly directly through reports of traveling government agents, entrepreneurs, and even spies, during the second half of the century Germany rapidly achieved technological preeminence and creativity. Technological innovations, such as the otto and diesel engines, now moved from Germany across the Channel.[2] The spread of the industrial revolution was connected with the rise of a system of higher technical education and the growth of a new profession: engineering.

Although the German university system was undergoing reforms during the early nineteenth century, they did not include the accommodation of technical subjects, but were much more oriented toward the principles of German idealism and academic freedom. The new technical schools in the country were modeled on the French Ecole Polytechnique, which had been founded in 1794 and was based on the systematic study of the natural sciences and mathematics as well as their application to technical problems. The first technological institute in German-speaking territory was founded at Vienna in 1815, the first in Germany proper at Karlsruhe in 1825. By 1836 almost all German states had higher technical schools that were modeled on Vienna or Karlsruhe. By the end of the century they were all officially called Technische Hochschulen. Not until 1899, and then by special decree of Kaiser Wilhelm II, however, were they made officially equal in degree-granting power to the older universities.[3]

The Technische Hochschulen produced among their faculties and students a network of teachers, entrepreneurs, engineers, and inventors who not only influenced each other in terms of technical subjects but also undoubtedly in terms of attitudes toward the interrelationship of technology and culture. For example, F. J. Redtenbacher, professor of mechanical engineering and founder of the science of machine design in Germany, taught at Karlsruhe from 1814 to 1863. Among his pupils were Eugen Langen, the industrialist who eventually teamed with Nicholas Otto to develop the otto internal combustion engine; Heinrich Buz, later director of the Augsburg Engine Works, the factory in which the diesel engine was developed; and Franz Reuleaux, subsequently professor at the

Berlin Technische Hochschule and exponent of an industrial decentralization theory that may well have influenced Diesel.

After the founding in 1855 of the polytechnical institute in Zurich, a number of prominent men gathered on its faculty: Reuleaux; Rudolf Clausius, professor of physics and one of the founders of modern thermodynamics; and Gustav Anton Zeuner, professor of mechanical engineering and contributor to thermodynamic theory. Carl Linde, founder of modern refrigeration methods, studied under Clausius, Zeuner, and Reuleaux. During the 1870s Diesel studied at Munich under Linde.[4]

The Technische Hochschulen, which combined theory and practice, are often considered to be one of the leading factors in Germany's rapid rise to technological prominence.[5] By the middle of the nineteenth century, they were producing professional engineers. The engineer had, of course, predated the nineteenth century, but, especially in that century and in Germany, the profession became consolidated through systematic education and the awarding of degrees. As one of the earliest German observers to write a critical work on engineers remarked, the modern engineering profession is bound up largely with concepts of energy that evolved especially after the invention of the steam engine.[6] In 1856 the Verein Deutscher Ingenieure (VDI), or German Engineers' Association, was founded to represent engineering interests.[7]

The rise of the Technische Hochschulen and the professional engineers was, however, not immediately accompanied by a corresponding rise in social position and prestige. Indeed, the proponents of the Technische Hochschulen fought throughout the nineteenth century for equal recognition with the universities and were rewarded with success only by Wilhelm II's direct intervention. To a large extent, the fight for equality with the universities in degree-granting power was also a battle for social equality of the engineering profession. In part, defenders of the universities looked down on the upstart technical schools, but in part, also, the situation was worsened by German idealism and romanticism. They tended to divide human activity between "higher" cultural pursuits, such as art, religion, and philosophy, and "lower" civilization, which encompassed politics, economics, and technology.[8] Technology in this view was strongly suspected of possibly leading to the destruction of culture. Later, at the end of the nineteenth and during the twentieth century, especially under the influence of the social dislocation caused by rapid industrialization, such attitudes would lead to a whole genre of antitechnological cultural criticism.[9]

Similar difficulties were encountered by the German engineer in his struggle for social acceptance. During the nineteenth century, he found himself looked down upon by the older, more prestigious professions, especially the lawyers, who monopolized governmental positions and who became one of the chief targets of engineers' complaints. Such organizations as the VDI were founded in part to represent the class interests of engineers. Complaints about lack of status became a common theme of engineering literature from the late nineteenth century on. Indeed, they can still be heard today—and not only in Germany.[10]

Although the struggle for acceptance was not confined to any one group of engineers, recent research suggests that mechanical engineers particularly may have been most adamant about lack of status. At the beginning of the nineteenth century, they tended to come from lower social orders, such as the artisans, and to possess less formal education than, for example, civil engineers. By the end of the century, differences between mechanical and other types of engineers were less apparent.[11]

One of the most perceptive and influential writers on the subject of engineers' status was Max Maria von Weber, eldest son of the composer Karl Maria von Weber. Von Weber had been trained at the Technical Institute in Dresden and at the University of Berlin. He became an expert in railroads and worked for the Saxon, Austrian, and Prussian railroad systems. He was a prolific writer of both technical and nontechnical works. In several publications during the 1870s, he discussed the situation of the engineer.[12] According to him, just as there can be in society individuals who are considered "upstarts" and "gatecrashers," so in the Germany of his day the entire professional class of engineers filled these roles. Von Weber bemoaned the fact that, because of the fairly recent appearance of this class, it was looked down upon by the older professions, such as law and medicine.

Part of the problem was that German technology was beginning to advance just as the nation was experiencing its greatest cultural outpouring in the period of idealism and romanticism. For this reason, technology was bound to be unfavorably compared to cultural pursuits and judged to be similar to handicrafts. Technological decisions were often still being made by dilettantes and lawyers. In contrast to England and France, where engineers were much more integrated into their societies, "there was still no glory for the German engineer."

According to Von Weber, the answer was the growth of a strong corporative spirit among engineers and cultivation of the traits of a "gentleman," which would allow the profession access to the best society. In an interesting comment, he explained that the latter goal was somewhat difficult for engineers because they were forced by the nature of their work to deal with classes of people of a lower social order and were therefore forced to descend to that level of thinking and speaking. In conclusion, he pointed in passing to a justification of the engineer's claims to higher social status that would commonly be found in later literature: the fact that the engineer not only possessed knowledge and ability, but that he also had a higher cultural mission.

A number of significant themes emerge from these ideas that were characteristic of the social and political stance of German engineers: the desire for prestige; a hostile attitude toward the older legal profession, which was thought to monopolize positions of power and prestige in society; and a sense of disdain for the working class. The solution emphasized the building of a corporative spirit and fostering the gentlemanly arts in order to be more easily accepted by the leading social classes. The justification of the engineers' claim to status lay in the cultural mission they performed. Despite their feelings toward the legal profession, they were obviously looking to the older, higher social classes as a model to emulate, rejecting any identification with lower classes as well as radicalism and assuming a connection between technological and cultural progress.

Similar ideas were represented in the leading engineering organization, the VDI, whose membership consisted primarily of mechanical and metallurgical engineers.[13] Although its founding as a national organization to represent the common interests of engineers clearly had political overtones, the organization from the first understood its mission as one of furthering professional interests, particularly by emphasizing the ability of technological advances alone to solve social problems. During its early years, the leadership contained a large percentage of entrepreneurs. Indeed, the German sociologist Gerd Hortleder has argued that, because engineers were drawn from such a heterogeneous background and lacked a common model to emulate, the entrepreneur was the only social type with which they could identify.[14]

It is not surprising that in their efforts to advance their own class, the engineers would deliberately distance themselves from the very group of craftsmen with whom they were often identified by their social superiors. Hence, their constant rejection of political radicalism, especially Marxism. The search for prestige almost automatically ruled out radical responses.

Even after the 1870s, when the engineer's role increasingly changed from independent entrepreneur to functionary of a large concern and fear was expressed about the proletarianization of the engineer,[15] the response of the VDI and like-minded engineers tended to focus on the mediating effect that the profession would exert on the antagonism between capital and labor. Not until the twentieth century, and increasingly during and after World War I, did engineers turn toward more political, especially technocratic, conceptions.[16]

Hortleder's thesis is that German engineers, especially as represented by the VDI throughout the nineteenth and into the twentieth century, have constantly thought in terms of professional advancement and in the ameliorating role of technology in social and political conflict, but rarely have they ever come to grips realistically with the social and political process and how to become involved with or change it. They have opted for a policy of "neutralization," that is, they saw (and see) themselves as a neutral mediating force. But, in so doing, they have essentially removed their profession from political and social conflicts.[17]

Representative of such ideas is Ludwig Brinkmann. At the end of his work *Der Ingenieur* (1908), he pointed to the growing tendency for engineers to become anonymous cogs in large establishments. He saw this tendency as one of the chief reasons for the growth of engineering organizations and their fight for such things as honorary titles, even for the same kind of social welfare legislation that was extended to workers. Brinkmann was suspicious of these efforts, however, for he thought that the engineer, in adopting the structure and goals of mass organizations, was reducing himself to the level of the worker. The engineer should understand that his problems were caused by the age of transition in which he lived.[18] In the coming age, all areas of human endeavor would be ruled by technology, and the profession would come into its own. "Then the engineer will cease to be a servant and will become the master."[19] Presumably, technological progress would somehow automatically solve the engineer's and society's problems.

Historian Wilhelm Treue has argued that the position taken by the German engineers was basically contradictory. As bearers of technological progress, which in turn would lead to political and social advancement, they desired a position of leadership in society. At the same time, they rejected unionism as well as socialism and desired acceptance by and position in an essentially conservative social order that begrudged them equal recognition. Their position, says Treue, was similar to the revisionist Social Democrats, who were attempting to initiate a nonbourgeois society by collaborating with that society.[20]

A similar phenomenon can perhaps be seen in the so-called "feudalization of the bourgeoisie." According to this theory, rather than articulating an independent political style, the upper middle class in imperial Germany increasingly aped the manners and lifestyles of the nobility and joined in their hostility toward the working class. This represented a tendency to combine economic progress with the desire for recognition by the older, ruling class.[21]

The foregoing comments on engineers' attempts to integrate into their society are not meant to suggest that they lacked any conception of or response to social problems of their day. Indeed, Wilhelm II himself had expressly stated that the Technische Hochschulen had a social and political task to perform. At the time he conferred doctoral-degree-granting powers on the Berlin-Charlottenberg school in 1899, he observed:

> I wanted to bring the Technische Hochschulen into the foreground, for they have great problems to solve—not only technical but also social. To date, these problems haven't been solved in the way I had wished. They [the TH's] can exert great influence on social relationships, since their connections with industry and labor permit a great deal of discussion and action.[22]

During the nineteenth century, in fact, much discussion of social problems occurred in technical circles.[23] Industrialists as well as engineers showed growing concern for the social question, and some began to advocate and carry out welfare measures for their workers. In many cases, they were motivated more by conservative, patriarchal principles than by perceived connections between technological and cultural advance. During the early nineteenth century, entrepreneurs and engineers were often the same people, or at least numerous successful entrepreneurs began as engineers and shared the same Technische Hochschule education. One of the best known examples is Alfred Krupp, who introduced a variety of funds for sickness and accidents as well as housing, schools, hospitals, canteens, even churches and playgrounds—while at the same time demanding absolute obedience from his workers.[24]

Another example of a conservative, patriarchal approach is Heinrich von Buz (1833–1918), director of the Augsburg Engine Works and Diesel's collaborator in the development of the diesel engine. Often referred to as the "Bismarck of German industry," Buz was ennobled for his accomplishments in 1907. He introduced a variety of welfare schemes from the 1870s through the first decade of the twentieth century, including reduction of work hours, accident insurance, and workers' housing. His primary motive was a hatred of unionism and Marxism.[25] Entrepreneurs such as Krupp and Buz anticipated Bismarck's social legislation of the 1880s, and for the same basic reason: to take the wind out of the socialists' sails by providing the workers with a variety of social welfare measures.

Besides conservative, paternalistic reactions, a number of businessmen and engineers responded to social problems based on their perception of technology's role in determining cultural progress. The essential arguments used by most of these men were that cultural progress was based on technological progress; that continuing technological advancement would ultimately solve problems that had originated in earlier technical advances; and that engineers, by virtue of their middle position between capital and labor, would ameliorate or lessen class antagonisms. Such ideas were generalized in the nineteenth century in many countries and accompanied the rise of the industrial revolution and technology.

In Germany, they were directly connected with the growth of the Technische Hochschulen, whose proponents attempted to counter their critics by emphasizing the cultural importance of technology. For example, Ferdinand Redtenbacher saw a link between technological gains and freedom. Technology was necessary for progress from the agrarian to the industrial state, the latter being judged as superior. The Technische Hochschulen should teach engineers humanistic subjects, such as philosophy, history, and languages. Redtenbacher optimistically saw the world as built on an unbroken chain of laws, founded on mathematics. The facts of social life no less than nature could be understood in this way.[26]

Such optimistic opinions about the relationship of technology and culture continued to be expounded by Technische Hochschulen teachers throughout the nineteenth and into the twentieth century. Typical are the works of the mechanical engineer Alois Riedler (1850–1936), who was both professor and ultimately rector of the Technische Hochschule at Berlin-Charlottenberg. He wrote extensively on the reform of technical education and was to be one of the chief critics of Diesel's originality. According to him, technology freed men from oppressive work and helped them win the "struggle for existence"; hence, it was the foundation of higher culture. The work of the engineer was critical because through it technology improved the lot of the mass of humanity and solved the most serious social problem, that of equality. Riedler pointed to such results of technology as the transportation revolution and the growth of a global civilization. He acknowledged that technological advance caused problems, but contended that the main difficulty was an economic organization that put profits and large-scale industry before the welfare of the general public. The state needed to invervene if private economic interests were damaging the public good. Riedler also bemoaned the limited recognition of the engineer or his profession.[27] He thus took the argument one step further by saying that progress was not automatic and that active steps might be necessary to redress wrongs, particularly in the economic organization of society.[28]

The ideas expounded by academics connected with the Technische Hochschulen were shared by a large number of entrepreneurs and engineers who had been trained in these institutions. Without a detailed study, it is difficult to say how widespread such opinions were, but they were held by several influential men. Behind the thought and actions of these individuals was the hope of winning the workers away from socialism. Hence, industrialists like Friedrich Harkort advocated forms of profit-sharing. Werner von Siemens, Ernst Abbe, head since 1889 of the Zeiss optical firm, and Wilhelm Ochelhäuser, chief executive of the Continental Gas Company (Continental-Gasgesellschaft) from 1856 to 1902 and a National Liberal Reichstag delegate, formulated a number of schemes for accident and sickness insurance. Siemens established workers' settlements, hospitals, libraries, and sports clubs. Siemens, Ochelhäuser, and Harkort believed that the continued progress of science and technology solved social problems by filling the wants of mankind, thus ameliorating class conflict and achieving the goals of the socialists without recourse to violent revolution or social change. To help further scientific advance, Siemens donated land in Berlin-Charlottenberg on which the Physikalisch-Technische Reichsanstalt was established. Otto Lilienthal, the aeronautical pioneer, thought that flight would destroy national boundaries, bring people together, even create a world speech and make war impossible.[29]

Both Siemens and Lilienthal contended that the spread of power would help decentralize industry and rescue the artisan from the growth of large factories. Indeed, this attitude was widely voiced at the end of the nineteenth century. It was usually tied to the development of a small power source, whether steam, gas, oil, or electric, which would be cheap to buy and economical to operate. These proposals were made by economists such as Gustav Schmoller and were frequently discussed in books and in the pages of engineers' journals like the *VDI-Zeitschrift*.[30] Both Otto and Langen believed that their engine was the salvation of the small craftsman.[31]

One of the most interesting examples of this industrial decentralization theory was propounded by Franz Reuleaux (1829–1905). A student of Redtenbacher at Karlsruhe, he taught machine design at Zurich and then at the Berlin Technische Hoschschule. He "was recognized within his own lifetime as a chief authority in mechanical engineering subjects associated with machine design," especially because of his influential *Lehrbuch der Kinematik*. He was involved with international exhibitions during the latter nineteenth century, fostered improvement of German industry, and served on the "research team" that developed the otto engine. He was also active in cultural affairs; for example, he translated Longfellow's *Hiawatha* into German and wrote a number of works on culture and technology.[32]

Reuleaux argued that it was essentially the steam engine that inaugurated the modern industrial age in advanced countries.[33] Not only had a virtually unlimited power source been introduced, but it had also led to mass manufacturing, in which utilitarian goods far surpassed the older artistic, handmade goods. Reuleaux's expression for the industrial form of organization was "machinofacture" as opposed to "manufacture," as in its older definition of "making by hand." In terms reminiscent of the young Karl Marx, he contended that the concentration of workers in factories caused the loss of individuality and led them more and more to being reduced to attendants of a machine.[34] The machine, more and more independent and capable of "intelligence," was practically replacing man. "However, man, its servant—horrible irony—sinks to the level of a machine."[35] From these circumstances emerged monotony, poor wages, and loss of contact with family life. These negative consequences of modern industry had produced the social question, "the burning question of our time."[36]

Reuleaux rejected the model of harmony presented by classical economics whereby competing individual forces would work for a greater good, though he stated that the effect of machines was not all bad. The answer was to be found in the small power source that would at least allow some craftsmen, such as weavers and woodworkers, to become economically competitive again. The steam engine was characterized by a tendency to large-scale manufacture; the larger it was, the more cheaply it could work. A power source that was inexpensive to operate and maintain could break the monopoly of large capital investments necessary for the steam engine:

The small master will become economically competitive in spite of certain advantages of large-scale industry, because in home work the mutual support of family members, that is, the moral element will enter as an important factor. In the environment of helpers and apprentices, the small master will build a closed work organism with head and members, with superior and subordinate power. It will be similar to the older craft manufactures, but different also because of the new power sources.[37]

Reuleaux pointed especially to the air and gas engines as well as to the newly tested oil engines. These were, he said, the "true power engines of the people."[38] Small- and large-scale industry together would produce a socially peaceful condition. "Thus, machine science can take a position in relation to the workers' question."[39] It is not surprising that, in other writings, Reuleaux proposed a corporative restructuring of society as a whole.[40]

Reuleaux's ideas were a combination of social criticism of the effects of large-scale industry with rather conservative, organic, corporative solutions to these problems.[41] Significant also is the idea that, if technology had caused social problems, its advance would solve them—apparently without causing additional ones.[42] Reuleaux's arguments for the small power source are directly related to the social ideas of Rudolf Diesel.

### Rudolf Diesel's Social Ideas

Diesel's approach to the society of his age was complex and ambivalent. For example, he very much wanted to be a success and to be accepted by society. Through his inventive success, he hoped to gain substantial profit. Indeed, in 1898, he was negotiating concerning his Russian patent with Emanuel Nobel, nephew of the discoverer of dynamite, Alfred Nobel, and owner of a Russian engine company and large oil fields in Baku. Diesel entertained grandiose plans at the time and even dreamed of competing with Rockefeller. "Perhaps I—one small man—will succeed in doing what all governments together have failed to do: throw out the Rockefellers—that would be amusing!"[43] Diesel's desire to become rich, famous, and socially accepted undoubtedly stems in large part from a reaction to the difficult circumstances of his early life.

These trying times perhaps account for Diesel's longtime interest in the social question. Although his son Eugen thinks this concern may have gone back to his student days in Munich and his apprenticeship with Sulzer in 1879 and 1880, it is clear that at least from the 1880s on, he was, alongside his inventive efforts, showing a growing interest in social problems.[44] In a letter to his parents on June 22, 1880, for the first time he mentioned his interest in the working class. This came before any indication in his letters of a desire to invent a new engine. After expressing a sense of alienation and a wish soon to leave France, he exclaimed: "I have no rights with regard to the state, cannot vote, and cannot take part in public life, which is one of my life goals, *since the workers' question is one close to my heart."* He continued, however, by saying that his attempts to be easygoing with the workers had totally failed: "One must be strict and put the fear of God into them, otherwise these people have no discipline."[45]

Diesel's condescending attitude toward the workers can also be seen in a document he prepared in the late 1880s entitled "Leading Principles of Factory Organization." It reads like an anticipation of both Taylorism and the expressionist silent film *Metropolis.* Diesel stressed the division of labor, order, and precision, as he said, "to [the point of] pedantry." The general director's office was to be in a corner where all entrances and exits could be observed. All communication apparatus was to be centralized in the office. A large bell would remind foremen of appointments with the director. Diesel argued that heating and lighting had to be adequate, even if it meant going against the wishes of the workers.[46] Here again, his concern for the workers is coupled with his disdain for their feelings. Just as with his engine, the most rational solution had to be adopted, a solution he was sure he knew. In view of this attitude, it is probably not surprising that the workers showed so little interest in his solutions to social problems.

Diesel's familiarity with factories throughout Europe may have stimulated his interest in the workers and how to help them.[47] His experiences during his Paris days also throw light on his growing social concerns: "He had 100 frc. a month, 30 for a room and 2 a day for food. He ate in workers' locales, on holidays for 22 sous, and with all of it felt quite well. His social ideas may well have begun during this period."[48] According to Eugen Diesel, his father read

Zola's novels.[49] As will be discussed later, he undoubtedly came into contact with the literature of the French utopian socialists during the 1880s or later.

Diesel continued his interest in the social question throughout the 1890s and the first decade of the twentieth century. Even when the family was on one of its many vacation trips, he would point out the slum areas in cities and call attention to the condition of the poor.[50] When he was negotiating with Emanuel Nobel in 1898 and obtained 800,000 marks for his Russian patent, he apparently considered setting up a special foundation to study social problems, thereby complementing Alfred Nobel's endowment for the Nobel prizes.[51] Indeed, Diesel himself apparently hoped to receive the Nobel prize, whose financial award would serve as the "founding capital for the solution of my great social problem."[52]

His social concerns were highlighted in *Solidarismus* (1903), which will be discussed later in this chapter. After its publication, he stated: "That I have invented the diesel engine is well and good, but my chief accomplishment is to have solved the social problem."[53] This quote not only illustrates his dual goals, it perhaps also illustrates a naiveté not uncommon among engineers. In any event, his interest in social matters continued. In 1907 he gave a well-received speech on his social ideals before a Berlin engineering society. He reported to his wife, "The engineer is entering a new era, in which he will gradually grasp the decisive influence in the state away from the lawyers."[54]

Essentially, Diesel's efforts to solve social problems involved two areas. First, his inventive endeavors were partially stimulated by the desire to find a small power source that would dethrone the steam engine and help the small artisan stay competitive.[55] That Diesel should be familiar with the industrial decentralization theory is not surprising because it was widely discussed, written about, and advocated during the latter part of the nineteenth century in Germany. Indeed, possibly Diesel's teacher Carl Linde made him aware of this idea because Linde was a pupil of Franz Reuleaux.[56]

The theme of the small power source can be traced through Diesel's writings into the first decade of the twentieth century. In May 1887, as he began constructing and testing his ammonia engine, he prepared a four-page list of uses for such a small engine. It included dentistry, jewelry making, weaving, woodworking, printing, and numerous other applications, such as in household tasks, restaurants, small boats, water pumps, and hospitals.[57]

Diesel's interest in the small power source also appeared during the period of his invention of the diesel engine. In the handwritten draft of his first patent application, dated 26 February 1892, he emphasized the small size of his proposed engine, as well as its independence from gas, electrical, or water resources. He went on to say: "All these characteristics make it especially well suited as a small engine. In this way a decentralization of craft industry is possible, instead of its accumulation and concentration in large cities. Such a process would involve political, economic, and hygienic advantages."[58]

Similar concerns were expressed in Diesel's first book, *Theorie und Konstruktion eines rationellen Wärmemotors,* published in 1893. Near the end of it, in a section entitled "Applications of the Engine," he envisioned the new engine as small, cheap, and light, comparable to a sewing machine. Thus, it could compete successfully with centralized power, such as gas:

> Therein lies a great advantage. It is undoubtedly better to decentralize small industry as much as possible and to try to get it established in the surroundings of the city, even in the countryside, instead of centralizing it in large cities where it is crowded together without air, light, or space. This goal can only be achieved by an independent machine, the one proposed here, which is easy to service. Undoubtedly the new engine can give a sounder development to small industry than recent trends which are false on economic, political, humanitarian, and hygienic grounds.[59]

Diesel also foresaw the application of his engine to locomotives, street cars, and other vehicles, which would help to decentralize transportation.[60]

Near the end of his June 1897 speech to the VDI annual meeting in Kassel, in which he announced the successful tests of the diesel engine, Diesel once again returned to the idea of decentralization of industry as a consequence of the engine.[61] As late as 1901, diesel promotional literature stressed, among other things, the engine's role in decentralization.[62] Furthermore, petitions of 1898, 1900, and 1904, directed by M.A.N. and Krupp to the German Bundesrat and aimed at lowering the tariff on imported oil, all stressed that the diesel engine was a small power source and that it was "one of the most important means to the solution of the social question."[63]

Thus, clearly Diesel saw his engine as a solution to small industry's problems. It is interesting that he came back to this theme during the period of innovation, even if at the same time he played up fuel economy or the particular advantages of the engine that would most likely interest a potential client. In fact, the first engines were not small, but were quite large and bulky (the 1897 engine was more than ten feet in height); and they tended, because of technological limitations, to be restricted to fairly large horsepowers.[64]

Diesel, Reuleaux, and other proponents of the decentralization theory were undoubtedly sincere, but to what extent were they correct? As the historian of the otto engine, Gustav Goldbeck, has suggested, proponents of the small power source were essentially backward-looking romantics. The fate of domestic industry was sealed, not by the lack of a small power source, but by the economic advantages of large-scale industry. No invention was going to turn back the clock to an earlier industrial age. Furthermore, insofar as industry did make use of small power sources, the steam engine was admirably suited for this role, despite what its critics thought its social effects were. For example, according to Goldbeck, in 1878 some 44,447 steam engines were operating in Germany, about 29 percent of which were less than 5 hp and 42 percent of which were between 5 and 20 hp. Diesel himself seemed unaware of the steam engine or the otto engine as a potential small power source and appears to have accepted the argument of the adherents of decentralization that steam power facilitated large-scale industry and concentration of the workers. Ultimately, if any power source became common to small industry after 1900, it was the electric motor. Proponents of gas and diesel engines as the small power source of the future eventually lost the battle to electricity.[65] Although the advocates of decentralization were wrong, they are an interesting example of social influences on technology and show that social motives can help bring forth new technologies.[66]

Diesel's second concern with the social question revolved around the organization of work. During the early years of the twentieth century, he began to collect statistics and other material on the economic situation in Germany. As he had solved the problem of an efficient heat engine, so too he would now solve the problem of the economic organization of society. And, as he had prefaced his inventive efforts with a written justification, *Theorie und Konstruktion,* so now he provided a written explanation of his social ideas: *Solidarismus.* Both books were meant to construct rational systems that must logically convince all who read them and ensure the success of their ideas.[67]

Indeed, Diesel used similar words in describing both his working process and *Solidarismus.* For example, in a letter to Carl Linde of February 11, 1892, the first public communication of his deliberations on the diesel engine, he said: "The results are not a conjecture or hope, they can be proven mathematically to the point that no doubt can exist that they are achievable."[68] In *Solidarismus,* he said: "You have seen that nothing in *Solidarismus* is doctrine, theory, arbitrariness or self-deception. Everything is developed and calculated strictly logically and numerically from life and from facts. Everything . . . is practically achievable."[69] Once one was convinced that everything in the book was true, said Diesel, he could have no excuse for refusing to cooperate in this great work of mankind.[70]

Diesel consciously used the German word *Solidarismus* (solidarism) instead of *Solidarität* (solidarity), which he thought referred to a sense of communal bonds or solidaristic interests. Solidarism, rather, was organized, conscious love of humanity; it was solidarity transformed into action. Solidarism was the understanding that the well-being of the individual was identical to that of the community. If the community should intervene to help the individual, the individual must understand that he needed to work and sacrifice for the community.[71] The title page of *Solidarismus* bears a large letter S that forms the center of a six-pointed star. Diesel explained that the six rays of the star represented the six main tenets of solidarism: veracity, justice, brotherhood, peace, compassion, and—the main tenet—love.[72]

In an attempt to provide a theoretical basis for his concept of solidarity, Diesel considered writing a book on natural religion. All that remains of this project is a handwritten outline, conceived about the year 1903, when he was working on *Solidarismus*. The manuscript on natural religion begins with the idea that the religious drive in man is really nothing more than fear of the unchangeable laws of nature. Gods were nothing more than personifications of natural laws, and the Christian religion had only attributed the powers of nature to one god, rather than to many. Now that men lived in a more enlightened, scientific age, it was possible to establish a more undogmatic natural religion which, of course, would change as man's knowledge increased. An example of a modern way to view an age-old religious idea was to see the concept of eternal life as an anticipation of the scientific laws of the eternality of matter and the conservation of energy.[73]

The logical conclusion of a natural religion, in which all laws affect all human beings equally, was common action to utilize natural laws when they were to man's advantage and, similarly, to avoid their effects when they were disadvantageous. This common action led to solidarism. The individual needed to realize that he could only be happy when the community was happy. As Diesel expressed it, solidarism was nothing more than the transformation of

Christian personal morality into the social, economic sphere. Instead of individual happiness in the hereafter, the emphasis now would be on the happiness of the community in this life.[74]

Thus, Diesel derived an ethical law from the laws of nature and provided a justification of his more concrete economic proposals. He even laid out the elements of a natural religion: foundation =unchangeable natural laws; articles of faith=the good of the individual depends on the good of the community, and active love of men; prayer=praise of altruistic love; sermons=teaching natural laws and how they support solidarism; and sacrifice=daily small sacrifice for the community.[75]

As with the predecessors of the diesel engine, the sources of Diesel's solidarism cannot always be easily pinpointed. His ideas, however, show distinct similarities with other intellectual movements of his time. In unpublished notes, he mentioned a number of authors and movements that expounded similar theories, though he always insisted his ideas were quite original.

One major source appears to be the French philosophers and utopian socialist thinkers of the nineteenth century, such as Claude Henri, Comte de St. Simon, François Fourier, Pierre Joseph Proudhon, Louis Blanc, and Auguste Comte. For example, Diesel's natural religion bears a strong resemblance to the ideas of St. Simon that all religions were codifications of scientific knowledge and that the world should accept a new Christianity, founded on the truths of positive science and brotherly love.[76] The idea of religion expressing scientific truths in a naive way was picked up by St. Simon from Enlightenment writers. Diesel may also have been influenced by Comte's ideas that society had progressed from the theological to the positive stage and that a new religion of humanity would preach love as well as the unity of mankind.

Eugen Diesel saw the similarity between his father's ideas and those of the French utopians. In a letter to his mother of June 2, 1916, when it appeared that she was going to write a biography of her late husband, Eugen remarked: "Do you know whether papa studied the two French philosophers, St. Simon and Comte? This must be firmly established for the biography, since the two have many ideas which papa expressed in almost the same manner."[77]

A second source of Diesel's social ideas is the general mid-to-late nineteenth-century enthusiasm for science, technology, and progress.[78] Although much of his terminology, such as "brothers" and "salvation," is borrowed from Christianity, by his college days, he had given up organized religion and was advocating tolerance, love of humanity, and social service—a kind of secularized social gospel. The idea of service to mankind in this life clearly influenced solidarism.[79] At one point, Diesel remarked that "Mankind is not bad . . . only badly governed."[80] He identified himself on the title page as "Engineer in Munich," and, indeed, Eugen Diesel was quite correct in indicating that *Solidarismus* was written in the firm conviction that the engineer, applying reason, could solve the pressing social problems of the time.[81]

Specifically, Diesel appears to have been influenced by the late-nineteenth-century version of philosophical monism, especially as propounded by the chemist and reformer Wilhelm Ostwald (1853–1932), with whom Diesel became acquainted during the first decade of the twentieth century. Based on the nineteenth-century science of thermodynamics, Ostwald held that energy was the substratum of all phenomena and that all change could be explained by the principles of the conservation of energy and of entropy.[82] In unpublished notes, Diesel referred to his idea of solidarism as "the last stone in the building of monism, the unified weltanschauung." As earlier ages had placed the earth or man's soul in the center of the universe, now "the progress of the whole is the goal of humanity."[83] Diesel and Ostwald had lively discussions concerning social problems, and in 1913 Ostwald asked Diesel to write an autobiography for a series Ostwald was editing called "Furtherers of Mankind." Nothing, however, came of this proposal.[84]

A third source for Diesel's ideas were the various solidaristic movements around the turn of the century. Solidarity was represented, for example, in the work of the Russian anarchist Prince

Peter Kropotkin, especially in his book *Mutual Aid* (1902), which emphasized the importance of cooperation as a basis for human affairs and as a much-needed corrective to the Darwinist idea of the struggle for existence. Diesel was aware of Kropotkin and corresponded with him in late 1903. In unpublished notes to *Solidarismus*, Diesel on several occasions mentioned Leo Tolstoy, especially his novel *Resurrection*, and quoted approvingly his condemnation of modern society.[85]

In the same notes, Diesel more than once referred to the American utopian writer Edward Bellamy, especially his last work, *Equality* (1897), which i .d been translated into German in 1898. Bellamy had spent the winter of 1868 and 1869 in Dresden and later said his experiences in Germany had opened his eyes to the social question. His works were quite popular in Germany, and his most famous book, *Looking Backward*, went through several German editions.[86] Diesel must have been influenced by Bellamy's concept of social solidarity, if not by his ideas of state socialism and the industrial army. Indeed, by Diesel's own admission, one sentence from *Solidarismus* is virtually lifted from Bellamy's *Equality*: "The individual should not do what the community can do better and more cheaply."[87]

Diesel was possibly also influenced by the German Catholic branch of solidarism, which taught that the unity of the individual with the community served the common good. The Jesuit economist Heinrich Pesch, for example, stressed the labor theory of value, social welfare, the harmony of labor and management, and the reconciliation of the individual with the well-being of society.[88]

Diesel appears to have been most directly affected by the French movement *Solidarité*, which was expounded by the sociologist Emile Durkheim, the philosopher Alfred Fouillé, and the politician Léon Bourgeois. Although Diesel claimed that his concept of solidarism was well formed before he became familiar with the French movement, his *Nachlass* contains numerous comments on a variety of French solidarist books. He had read, for example, Bourgeois's book *Solidarité* (1897) and had taken notes on it. He commented, "What I call solidarism is already called in Bourgeois's book 'movement solidarité.'"[89]

*Solidarité* became the semiofficial policy of the French Radical Socialist (liberal bourgeois) party from 1900 to 1914. Essentially, it was a philosophy that attempted to justify state intervention to redress social inqualities, while leaving the basis of capitalist society untouched. Viewed as the perfect compromise between socialism and laissez-faire, it proclaimed natural solidarity and justified social welfare, but rejected revolutionary solutions. According to Fouillé and Bourgeois, men were to act as if society were based on a contract, mutually agreed upon by its members. The result of this "quasi-contract" was a society analogous to an organism, based on the division of labor and governed by the idea of solidarity. Each member of society owed a debt to everyone else for services rendered. Solidarity was proclaimed as a scientific fact, not some vague sentiment. According to Fouillé and Bourgeois, the state could redress social wrongs, but it was not to interfere with the organization of the productive process. Much of this philosophy was meant expressly to combat the political and social ideas of social Darwinism.[90] Apparently, Diesel may have been especially influenced by the supposed scientific nature of solidarity and its contract theory, though not by the idea of state intervention.

Diesel's book *Solidarismus* is divided into two parts. The first is a theoretical justification and description of the principles of solidarism; the second is a painstaking outline of the exact relations and conditions in solidaristic contracts and arrangements, almost as if Diesel believed the more he could spell out the exact details of his plan, the more likely it was to be accepted. His concern in *Solidarismus* was economics. As he said elsewhere, this field was the most critical one for the freeing of the masses; it would pull politics along in its wake.[91] Indeed, he seemed much more concerned with describing the financial and administrative details of his solidaristic enterprises than in discussing exactly how they would work in practice.

In his book, Diesel propounded voluntary workers' action within the confines of the state. He adopted the commonly accepted nineteenth-century idea of the labor theory of value, but argued that in the future it would be cooperative work that would truly ennoble man and give value to the product of his work.[92] He said that 50 million workers in Germany were dependent on wages and salaries. If only they would unite and act together, especially in the simple form of contributing a penny a day to a "peoples' bank" (*Volkskasse*), they would soon be in command of huge sums of money that could be invested in secure state papers that would earn good interest. These funds, in turn, would be used by the peoples' bank as credit upon which cooperative industrial and agricultural enterprises could be founded. Because the workers could directly control their own enterprises and share in the profits, they would be motivated to work hard.

Echoing an analogy between perfect social organizations and social insects that goes back as far as the ancient Greeks, Diesel suggested calling his cooperatives "beehives" and the workers "bees," a choice of terms that the economist Lujo Brentano pointed out to Diesel was not calculated to win much sympathy from the workers.[93] Diesel logically carried out the analogy to such things as bee identification cards, bee contracts, and the rights as well as duties of bees. The "rationality" of it all prevented him from seeing its ludicrousness.

The peoples' bank and the cooperative factories were both based on contracts—among the contributing workers for the former and between the peoples' bank and the workers for the latter—that described rights, duties, and governing structure. Disputes were handled by arbitration; if a worker violated the contract, he was simply cut off from the privileges of membership.[94] The peoples' bank, which provided the element of coordination and control, was the ultimate owner of solidaristic enterprises and made sure that no competition existed between them and that no unneeded ones were constructed.[95]

The purpose of the coops appears to be twofold. First, they facilitated the construction of an exchange system, which existed within the regular economy. They produced goods at cheap prices and then exchanged them with each other. Each coop maintained an exchange store, in which workers could purchase products at cheap prices.[96] Rather than paying, individuals would carry cards that listed assets in the form of coupons that were canceled when purchases were made.[97] Second, the coops would provide a rather complete set of welfare measures for their workers. Diesel's examples were canteens, housing, a hospital that would provide full maternity and nursery facilities and be open to all without regard to marital status, medical care, sports facilities, convalescent homes, summer camps, nurseries, elementary and trade schools, libraries, lectures, and musical presentations. All these services were to be free to "bees" and their families.[98]

Most of Diesel's concrete proposals appear to reflect the French utopian socialists, though the exact manner in which he might have become acquainted with them is uncertain.[99] The creation of some form of worker cooperatives and the providing of cheap credit is involved in François Fourier's phalanx idea as well as in the social workshop concept of Louis Blanc, who also wanted the system of workshops to take over social welfare measures.[100] Pierre Joseph Proudhon contended that the notion of contract was the basis of society. He proposed workers' associations backed by a "peoples' bank," which guaranteed their enterprises, and even suggested a form of credit coupon that workers could exchange for the products of other workers.[101]

In all these thinkers, as well as St. Simon, the emphasis is on an attempt to overcome class antagonism through cooperative enterprises, not revolution. Terms such as mutualism or solidarity were often used to describe these proposals. The basis of Diesel's cooperative system was, of course, his concept of solidarism, which embraced the actions of the individual for the community and the intervention of the community for the individual. He maintained that solidarism was a scientifically proven fact and tried to show that society was already partially organized on solidaristic principles. Industrial examples that he cited were the social welfare measures of Krupp and of the chemical concerns Bayer and Badische Anilin-und Soda-Fabrik (BASF).[102]

Diesel emphasized the benefits of a solidaristic organization of society. It would stimulate a more active family life, enhance domestic values, and eliminate the "born criminal type" by removing the causes of social misery. It would solve the most serious modern economic problem: anarchistic production leading to periodic crises. It would allow for a decentralization of industry and the use of technology for human freedom, not slavery. It would abolish strikes as well as labor unrest and bring an end to class conflict, perhaps even international conflict, by preaching cooperation instead of antagonism.[103]

Despite solidarism's wide-ranging effects, it supposedly would not alter the basis of the state and its existing classes. In fact, it would not be achieved by the state but by individual action.[104] Diesel emphasized that no change would be required in social structure or laws. He appealed to all classes to support solidarism. It would not destroy, but would increase the middle class (*Mittelstand*). It would not threaten religion, but would expand its morality to encompass the social sphere.[105]

Although Diesel constantly emphasized the necessity of working within nineteenth-century German society and leaving existing social relations untouched, this may have been done to make his ideas more easily acceptable. In his unpublished notes on solidarism, his opinions were much harsher: "Judges and lawyers are greater criminals than those they have helped imprison . . . present conditions are unjust, unreasonable and cruel." Men and women are not by nature criminals or prostitutes, but only made so by social conditions. "Everything in our social organization is false and perverted."[106] One wonders whether or not Diesel actually saw solidarism as a process that would ultimately transform all of society.

Diesel had high expectations for his ideas. For a while, he continued to spend time and effort to further them. He gave several speeches on solidarism. For example, at a meeting of German consumer unions in June 1904, he emphasized that only an act of will was required to create the first cooperatives. Perhaps some philanthropist or the consumer unions themselves could be convinced to provide the funds to found the first "beehives." Diesel stressed that he did not believe mankind was too inert or "egotistical" to cooperate. He believed in the progress of humanity and in the force of ideas whose time had come. [107]

Diesel carried on correspondence and held conversations with various notables. Peter Kropotkin, in typical anarchist fashion, thought Diesel's ideas would require too much political centralization. Lujo Brentano suggested that he might win adherents, especially among the revisionist socialists, but that his tactics were incorrect. He ought to have emphasized that solidarity was nothing but a device by which the socialists could reach their goal. Others thought Diesel's view of human nature was too unrealistic, that the workers could not be expected to contribute regularly to something in which they saw no direct advantage, and that solidarism would destroy individual initiative. Diesel's former teacher Carl Linde argued that the very type of inventive effort Diesel had displayed in the creation of his own engine would cease if his solidaristic ideas were ever put into practice.

Diesel made lists of groups that might be of use in the cause of solidarism, such as the Social Democrats, unions, savings banks, and consumer groups. They should be invited to form a committee. People to approach included Friedrich Naumann, Lujo Brentano, philanthropists, and even foreign politicians, such as Joseph Chamberlain, in England, Alexandre Millerand and Charles de Freycinet, in France, and Theodore Roosevelt, in the United States. Among magazines that might be useful were *Feld und Garten* and the anti-Semitic Theodor Fritsch's *Hammer*. [108]

Published reaction ranged from ignoring *Solidarismus* to the observation that the book lacked any scientific merit and was "a real pain" to read to the end. The *Sozialistische Monatshefte*, organ of the revisionist, or evolutionary, socialists, dismissed the volume as "crass utopianism," and *Documente des Sozialismus* argued that widely diverse groups of people can only be made to cooperate by force.[109] Gradually, Diesel's disappointment increased. The workers did not rally to his cause. He turned to further work on the diesel engine. His health and financial problems grew, absorbing more and more of his time.

The one concrete result of Diesel's ideas may have been the inclusion in all of his diesel engine contracts of the so-called reciprocity, or solidarity, clause. By its terms, each diesel licensee was to share all improvements with all other licensees. Diesel idealistically referred to such clauses as "the fundamental principle of all diesel engine contracts."[110] As shall be pointed out in chapter 5, however, the principle of solidarity was widely violated, as each company looked out for its own interests. Some speculation has even been made that Diesel's growing interest in social conditions may have alienated him from industrialists who did not wish to work with a would-be social reformer.[111] In 1925 Oldenbourg Verlag recycled the last four hundred copies of the book because of lack of sales. *Solidarismus* remained "like so many of its predecessors, a printed utopia, a beautiful dream."[112]

Diesel exhibited many of the trends and characteristics common among engineers in nineteenth-century Germany. An outstanding graduate of the new Technische Hochschule system, he, like many others in his profession, sought acceptance and status in imperial German society. He, too, held technocratic ideas, sought increasing influence by engineers in the political process, and regarded the legal profession as the enemy. He believed in scientific as well as technological progress and felt technology could help solve the social problems of the time. Like Franz Reuleaux, he advocated a small power source—in this case his own engine—that would help decentralize and save the small craftsman from the economic competition of larger concerns.

His doctrine of solidarism may seem even more radical than some engineering solutions to social problems. Yet, it is essentially an economic utopia based on the old concept of an organic society. Such organic theories were held by other engineers who were concerned with social problems, such as Reuleaux, Carl von Bach, and Wichard von Moellendorff.[113] Solidarism is nonpolitical or politically neutral, rejecting class antagonism and emphasizing the moral influence of the family and a harmonious society. Diesel made much of the self-sufficiency of solidaristic enterprises and their effect of decentralizing the economy. Interestingly enough, however, in contrast to his earlier decentralization theories, he made little mention of technology in *Solidarismus* outside the vague assertion that it would be used for progressive purposes rather than for enslaving mankind.[114] Perhaps by 1903 the development of the diesel engine in large sizes and powers made Diesel de-emphasize his former theories on the small power source. Perhaps also, industrial decentralization had taken a back seat in his plans to a more general economic reorganization based on solidaristic principles.

Despite Diesel's sincere desire to cure society's ills through solidarism, an argument can be made that his ideas, as well as those of like-minded engineers, are really quite conservative. They share the contradictions mentioned by Treue of attempting to work within and maintain the very society they also wish to change. Their projected programs and solutions show little understanding of the political process or the competing forces in modern society. It is no wonder, then, that their social ideas have had such slight impact. The lack of realism indicates why many such engineering solutions do not work, in fact, why engineers have exercised so little direct control over how society has used their inventions.

It can also be argued that Diesel's ideas and the whole status-seeking conservative approach of late-nineteenth-century German engineers was part of the general trend, mentioned earlier in this chapter, called the "feudalization of the bourgeoisie." Consciously or not, in taking a nonpolitical stance and arguing for the basic maintenance of institutions, engineers were doing their part in maintaining class relationships and the power structure of imperial Germany.[115]

It may be that engineering, as a fairly recently established profession and one that views itself as closer to business than to labor, has naturally focused on achieving professional and social status and that technocratic claims to political leadership are just another way of advancing these status claims. Engineering thinking may also tend to follow mechanistic lines, which result in the construction of "frictionless" utopias. Diesel seemed especially prone to building logical systems that were supposed to succeed because of their very rationality. This abstract rationality, which was often divorced from reality, and his lack of concern for what the workers really thought, made practical implementation of his ideas impossible.

The concept of solidarity is still alive today and not only in the workers' movement in Poland. A recent article has suggested that it can be applied to United States politics.[116] The question of the social responsibility of engineers is also very much alive. If they wish to make an impact on the political and social process, they may be able to draw some lessons from the past endeavors of their colleagues.

# CHAPTER 3

## The Invention of the
## Diesel Engine

•

Toward the end of his life, Rudolf Diesel was to say of the inventive stage[1] that it was a "happy time," in which one first intuitively reached an idea, and then, struggling and compromising with reality, achieved a working model. Because only a small part of the ideal would ever be realized,

> one must desire a lot in order to reach anything . . . invention means guiding a correct idea, which has been sifted out from a large group of erroneous ideas, through numerous failures and compromises to practical success. For that reason, every inventor must be an optimist.[2]

This chapter discusses the "happy time," primarily in 1892 and 1893, during which Diesel worked out the theory of the diesel engine. Chapter 4 will examine the development stage, the struggle to construct the working engine. The main emphasis in the present chapter will be on the train of thought that led first to an unrealizable ideal and then to its modification to fit a practical combustion cycle. Attention will also be given to advances in thermodynamics and heat engines that formed the background of Diesel's ideas and to the effect of patent law on his inventive efforts.

## Internal Combustion Engines and Thermodynamics in the Nineteenth Century

The background to the invention of the diesel engine is the development of the heat engine, in particular the steam engine, since the latter part of the eighteenth century. A heat engine "converts heat into work by adding heat to a working fluid—usually a gas—so that the fluid expands and exerts pressure on a piston or on turbine blades."[3] Because, however, the steam engine was quite inefficient in terms of fuel economy, numerous attempts were made, especially from the middle of the nineteenth century on, to devise a more efficient engine, either steam or internal combustion. The latter engine is one in which the combustion that generates heat takes place inside the engine itself rather than in a special chamber. Simpler than the steam engine, the internal combustion type eliminates smokestack, boiler, and condenser. Another advantage is realization of higher pressures in the cylinder when air is used as a working medium; this results in more efficient use of fuel. Proponents hoped the internal combustion engine or improved small steam engine would be easier and more economical to use. This would help small craftsmen, who could not compete with the larger steam engines, which required large capital sums and big industrial units. This economic argument for a small power source is equivalent to the social argument that was discussed in chapter 2.[4]

One of the major difficulties impeding the perfection of the internal combustion engine was the problem of fuel. Coal, the primary fuel for the steam engine, was unsuitable for the internal combustion engine. A new and more suitable fuel, illuminating gas, was found in the early part of the nineteenth century, and by the 1850s numerous patents were being granted on a variety of gas engines. The most successful of these, before the otto engine, was that of Etienne Lenoir (1822–1900), who in 1860 introduced a double-acting engine that operated at atmospheric pressure. A mixture of illuminating gas and air was sucked into one end of the cylinder and ignited by an electric spark halfway through the stroke. The resulting explosion drove the piston to the end of the stroke. On the piston's way back, the burned gases were expelled from the cylinder by the piston. Meanwhile, a similar process in the other side of the cylinder and by the other side of the piston occurred. The two processes were phased 180 degrees of crankshaft apart so that each explosion on one side helped to expel the burned gases from the other side.[5] Between the 1860s and 1890s, hundreds of men were working on variations of the small heat engine, whether gas, oil, or hot air, both internal and external combustion.[6]

As the search for more efficient engines progressed and as more of the required technical hardware, such as pistons, cylinders, and valves, were perfected, the parallel development of the science of thermodynamics also continued throughout the nineteenth century, partly influencing and partly being influenced by practical engineering procedures.[7] Perhaps the most prominent figure in the founding of thermodynamics was the Frenchman Nicholas Leonard Sadi Carnot (1796–1832).

Carnot was a graduate of the Ecole Polytechnique. His father, Lazare, had been a member of the Directory during the revolutionary period and had served as minister of war under Napoleon. His nephew, Marie François, whom Diesel had seen during the Bastille Day celebrations in July 1888, was president of the French Third Republic from 1887 until his assassination in 1894.[8] Carnot's book, *Reflections on the Motive Power of Heat and on Machines Fitted to Develop This Power (Réflexions sur la puissance motrice du feu et sur les machines propres à développer cette puissance)*, appeared in 1824.[9]

Carnot's purpose in writing this volume was both to stimulate the wider use of the steam engine in France and to provide a general theory of heat engines.[10] This latter task was complicated because he did not understand the equivalence of heat and work as forms of energy nor did he know about the concept of absolute temperature.[11] Indeed, the modern laws of thermodynamics had not yet been elaborated. Yet, Carnot's genius allowed him to draw a number of extremely significant conclusions about the dynamic properties of heat as well as the working of heat engines.

In his day, analogies between heat and water engines were quite common and were strengthened by the generally held belief that heat was a substance or subtle fluid called the "caloric." Carnot envisioned heat engines working in much the same way that water engines worked: as water will flow from a high level or pressure to a low level or pressure, so too heat, or the caloric, will flow from a high level or hot body (boiler) through the engine to a low level or cold body (condenser). Carnot therefore stated that the power of an engine depended on the temperature differential through which the engine operated and not on the working medium. Practically speaking, however, he recognized that the medium was important because it affected such things as engine design and the temperature at which the engine could operate.

Carnot concluded that gases were the best working media because they expanded most for a given temperature change. He recognized the advantages of air as a working medium and suggested that it be compressed and then cooled as much as possible to allow for a greater temperature differential and, hence, motive power. He thereby foresaw the development of the internal combustion engine, which uses air both as a working medium and as an element in the combustion process, as well as the possibility of compression ignition, that is, igniting the fuel by compression alone.[12]

Carnot introduced the idea of a cycle in a heat engine, that is, a series of processes that restore a system to its original condition. Air in a cylinder, for example, can be compressed, heated, expanded, and cooled, and then brought back to its original state.[13] He was interested in describing the perfect engine cycle, which would generate the maximum amount of power from the fall of heat through an engine. In such a cycle, no useless flow of heat would occur. All heat generated would be turned into motive power or work. The engine could then be run in reverse, and the cycle would be restored to its pristine state and all original heat restored to it. In other words, motive power or work would be turned back into heat at the same temperature with which the cycle began. It is as if a water mill could generate enough power from the fall of water to pump the water back up to its original level, thereby producing a reversible process and a perfect engine. As Lynwood Bryant has stated, "No heat engine operating between given temperatures can possibly be more efficient than one operating on a cycle with all processes reversible."[14]

Carnot described what an ideal engine would look like, operating on his perfect cycle (see figure 1).[15] The ideal engine would consist of a cylinder containing a working medium, for example a gas, and fitted with a piston. In addition, there would be a hot body, from which the cylinder could draw heat, and a cold body, into which it could discharge heat. The cycle consists of four stages. First, the cylinder is brought into contact with the hot body. The working medium, whose temperature is close to that of the hot body, begins to expand but draws enough heat from the hot body to maintain a constant temperature. Although the temperature of the gas remains constant, its pressure will begin to fall in inverse ratio to its increase in volume. Second, the cylinder is isolated from the hot body. The working medium continues to expand, but its temperature begins to fall, its pressure dropping more rapidly than before. At the end of the second stage, the temperature of the working medium has fallen close to that of the cold body.

This is the end of the power stroke, during which the expansion of the working medium against the resistance of the piston is a conversion of heat into work. To complete the cycle, the piston must be driven back to its original position, and the working medium must be restored to its original temperature and pressure. This process will cost something: "negative work" by the piston on the working medium will be required to return the piston to its original position.

During the third stage, the cylinder is brought into contact with the cold body. The working medium is compressed, but its tem-

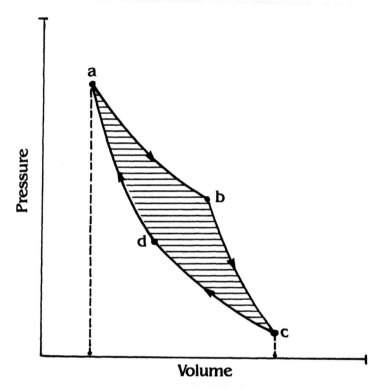

FIGURE 1. PRESSURE-VOLUME DIAGRAM OF THE CARNOT CYCLE.
Line a-b-c traces the first two stages, or power stroke, of the cycle, as the pressure
falls and the working medium expands. Line a-b traces the stage of isothermal
expansion, when the working medium is in contact with a hot body and is expanding with constant temperature. Line b-c traces the second stage of adiabatic
expansion, when the heat source is removed and the temperature of the working
medium drops. The area under the curve a-b-c represents the "positive work" of
the working medium on the piston. Line c-d-a traces the second two stages, or
compression stroke, of the cycle, as the volume decreases and the pressure rises.
Line c-d shows isothermal compression at a constant temperature, as the working
medium contracts but gives up heat to the cold body. Line d-a shows adiabatic
compression, as the working medium is isolated from the cold body and its temperature rises back to its initial value at the beginning of the cycle. The area
under c-d-a represents the "negative work" of the piston on the working medium.
The shaded area between the lines represents the "net" work available to the
engine, after one cycle is completed. Drawing by Ellen Thomson.

perature does not rise because it gives up enough heat to the cold body to maintain a constant temperature. The working medium's pressure begins to rise in inverse ratio to the decrease in volume. During the fourth stage, the cylinder is isolated from the cold body. The working medium is further compressed, but its temperature begins to rise, causing its pressure to climb more rapidly than before. When the temperature returns to its original value, the cycle is complete. Both working medium and engine have been restored to their original condition.

In modern terminology, the cycle's first two stages, or power stroke, consist of an isothermal expansion, in which the temperature is constant, and an adiabatic expansion, where no flow of heat in or out of the working medium occurs. The second two stages, or compression stroke, consist of an isothermal compression, in which the temperature is constant, and an adiabatic compression, in which again no flow of heat in or out of the working medium occurs. This is the ideal Carnot cycle, which is still studied today in every engineering course in thermodynamics.

The result of Carnot's theories was a first statement of what later became known as the second law of thermodynamics. Whenever energy is converted from one form into another, some loss will always occur through radiation or waste heat. Real engines are irreversible and always operate with such losses. Carnot's perfect, reversible engine would not suffer any losses due to friction or radiation. An engine that could end its cycle with more energy than it began would be a perpetual motion machine, and therefore an impossibility.[16]

At the end of his book, Carnot issued a warning that was forgotten by some of his successors, including Diesel:

> We should not expect ever to utilize in practice all the motive power of combustibles. The attempts made to attain this result would be far more hurtful than useful if they caused other important considerations to be neglected. The economy of the combustible is only one of the conditions to be fulfilled in heat engines. It should often

give precedence to safety, to strength, to the durability of the engine, to the small space which it must occupy, to small cost of installation, etc. To know how to appreciate in each case, at their true value, the considerations of convenience and economy which may present themselves; to know how to discern the more important of those which are only accessories; to balance them properly against each other, in order to attain the best results by the simplest means: such should be the leading characteristics of the man called to direct, to co-ordinate among themselves the labors of his comrades, to make them co-operate towards one useful end, of whatsoever sort it may be.[17]

As Professor S. S. Wilson has pointed out, Carnot was not proposing an attainable goal, but a "hypothetical cycle with which one could study thermodynamic concepts such as the maximum motive power of an engine."[18] Carnot's ideas were forgotten for a decade, until they were revived by the French engineer Emile Clapeyron (1799–1864), who first represented Carnot's cycle in terms of a Watt indicator diagram, and by the German physicist Rudolf Clausius (1822–88). Carnot's ideas were then widely disseminated; they strongly influenced Gustav Zeuner, author of the authoritative textbook *Technische Thermodynamik* (1887). Zeuner stressed the value of the Carnot cycle and held it up as the standard by which all heat engines should be judged.[19] Carl Linde, who studied under both Clausius and Zeuner at Zurich, transmitted Carnot's ideas to his pupil Rudolf Diesel.

In a letter of 1893 to Zeuner, Diesel stated, "Your book stimulated me to my work; . . . I have read it over and over. I might almost say my work is the direct consequence of the viewpoints expressed in your book. I have always thought of myself as your pupil."[20] Diesel's comments may be exaggerated, for at this time he was trying to obtain a testimonial from Zeuner concerning his new theory. Undoubtedly, however, Diesel was a strong admirer of Zeuner and was much influenced by him.[21]

By the 1850s the caloric theory of heat was giving way to the idea of heat and work as forms of energy. The laws of thermodynamics and the concept of absolute temperature were being formulated by such men as Clausius and William Thompson, Lord Kelvin. The advances in thermodynamics provided a new way of measuring a heat engine, namely by thermal efficiency, or the fraction of the total heat supplied to an engine that is turned into work in the cylinder. The modern term is "indicated thermal efficiency," which is defined as the work done by the gas in the cylinder divided by the heating value of the fuel input to the cylinder. By this definition, the thermal efficiency of a Carnot engine is 1 or 100 percent. Such a criterion could lead to evaluations that show a shockingly low thermal efficiency for the steam engine, perhaps around 7 percent. Professors such as Ferdinand Redtenbacher taught their students that there ought to be a better way to convert heat to work.[22] It was such ideas of thermal efficiency that Diesel would learn and that would provide the chief impetus for his own engine. Indeed, he pointed out steam engines to his family when they were in railroad stations and indignantly criticized their low efficiency.[23]

Several problems, however, were related to this new standard. As Lynwood Bryant has pointed out, "The change from the common-sense criterion of fuel economy to the new criterion of thermal efficiency is the first step into the domain of abstractions, of invisible things like heat and energy, where the footing is treacherous and common sense is not always a safe guide."[24] For one thing, critics charged that the steam engine was being subjected to an unfair comparison. It ought not to be held responsible for using all the heat energy in the fuel; rather, its efficiency ought to be measured against the *available* energy, which depends on the range of temperatures through which it is working. Based on the rather modest limits of temperature involved, the efficiency of the steam engine is not that bad.

Further, emphasizing thermal efficiency at the expense of the overall economy of a system can lead to distorted comparisons of engines. Thermal efficiency must be balanced against other efficiencies, such as brake thermal efficiency, which is the work output at the engine shaft divided by the chemical energy of the fuel input to the cylinder and which takes into account the loss of energy in an engine through internal frictions. A point is reached where gains in thermal efficiency are outweighed by losses in other efficiencies. Yet, "the conventional thermal efficiency remained the standard criterion."[25]

During the latter part of the nineteenth century, the search for an effective internal combustion engine was brought to fruition by Nicholas A. Otto, who perfected the "silent otto," the ancestor of today's internal combustion gasoline engine.[26] Otto and the industrialist Eugen Langen had founded the firm N. A. Otto & Cie. in 1864. During the 1860s Otto developed a gas internal combustion engine that operated at atmospheric pressure. Its fuel economy was good, but it was noisy and capable of producing only limited horsepower.

In 1872 the factory was reorganized as the Gasmotorenfabrik Deutz, located across the Rhine River from Cologne. Langen assembled what amounted to a research and development team unusual for its day: Franz Reuleaux as a consultant, Gottlieb Daimler (1834–1900) as general manager, and Wilhelm Maybach (1846–1929) as machine designer.[27] The silent otto, perfected by 1876, drew in a mixture of gas and air, "compressed it within the cylinder with a compression ratio of about 2½:1, and ignited it with a flame."[28] The thermal efficiency was about 14 percent, two or three times as good as a comparable steam engine.

The engine operated with a four-stroke cycle. The four-stroke combustion cycle in an otto engine consists of: (1) an intake stroke, in which a mixture of fresh air and fuel is drawn into the cylinder as the piston moves downward toward bottom dead center; (2) a compression stroke, in which the piston moves upward toward top dead center; (3) a power stroke, in which the fuel is ignited by some form of ignition device and drives the piston to the bottom of its stroke; and (4) an exhaust stroke, in which the piston moves upward and forces the burnt gases out of the cylinder. Otto's patents were broken in 1884, when it was discovered that the four-stroke cycle had first been mentioned in 1862 in an obscure pamphlet by the French engineer Alphonse Beau de Rochas.[29]

Otto was not an educated scientist or engineer. He had been a traveling salesman who heard about the Lenoir engine and became interested in the problem of controlling explosive fuel. Both he and Langen sought to provide a power source for small craftsmen, partly at least for the social motive of making them more competitive with large-scale steam production.

The success of the otto engine, however, did not decrease attempts by hundreds of inventors to experiment with a variety of working media in hopes of achieving higher efficiencies or constructing a variety of internal combustion engines.[30] Much concern still existed about the heat loss in gas engines because of the high temperatures and pressures generated through combustion and about the escape of unburnt gases at high temperature.[31] Although the Carnot cycle was judged to be unrealizable in practice, experts stressed the importance of approaching the ideal cycle as closely as possible.[32] Much uncertainty still remained as to what happened during combustion in a cylinder.[33] Despite the interrelationship of science and technology, theoreticians did not yet appreciate the difficulties of engine design, and engineers did not totally understand the thermodynamic underpinnings of their profession. It was in this milieu that Rudolf Diesel was educated and in which he undertook his inventive work.

### Diesel and the Rational Heat Engine

According to Diesel's own testimony, which is given in his 1913 *Die Entstehung des Dieselmotors*, written some time after he had invented the diesel engine, the original impetus for his invention came from his teacher Carl Linde during Diesel's school days at the Munich Technische Hochschule. Diesel heard Linde's lectures on theoretical machine design from the winter semester of 1877–78

Carl von Linde, inventor of
modern refrigeration techniques
and Rudolf Diesel's teacher.
(M.A.N. Werkarchiv)

through the winter of 1878–79.[34] (For a chronology of the invention and development of the diesel engine, 1878–97, see appendix 1.) During the course of these lectures, Linde discussed the by-then well-known Carnot cycle. Diesel's comment on Linde's explanation was as follows:

> When in 1878 my respected teacher, Professor Linde, explained to his audience in the thermodynamics lecture at the Munich polytechnicum that the steam engine transformed only 6–10 percent of the available heat [*der disponiblen Wärme*] of the fuel into effective work, and as he explained the Carnot theorem that in an isothermal change of state of a gas all added heat is converted into work, I wrote on the margin of my notebook: "Study whether it is not possible practically to realize the isotherm." Then I undertook to solve the problem! That was not yet an invention, not even the idea of one. The wish to realize the ideal Carnot process ruled my existence from then on. I left school, went into practice, [and] had to make a position for myself in life. [But] the idea followed me uninterruptedly.[35]

As Eugen Diesel relates in his biography, his father was quoting the marginal comment on the isotherm from memory because the notebook had disappeared despite Diesel's vain effort in 1912 to locate it.[36] The one comment on Carnot the author of this volume has found in looking at Diesel's college notes in the M.A.N. Werkarchiv is the following: "This book [Carnot's *Réflexions*] has chiefly historical value; it contains, however, an abundance of stimulating thoughts and gives a glimpse into the manner in which studies were pursued at that time."[37] It is not clear from the context whether Diesel was copying down Linde's comment or interjecting his own idea. The comment is not particularly enthusiastic, but at the same time is not very negative either. Other college notes that Eugen Diesel cites show that his father was stimulated to think about realizing a more ideal cycle in the steam engine or perhaps abandoning steam as a working medium completely and transforming the heat of coal directly into work. As the elder Diesel added, however, "how is that practically feasible? That is exactly what one must find out!!"[38] Apparently, then, Linde's comments must have stimulated him to think about the possibility of a more efficient engine than the steam engine. Furthermore, the idea of efficiency was linked with the Carnot cycle.

In 1913, near the end of his life, Diesel reflected on the role that the Carnot cycle had played in his era. Allowing for some after-the-fact exaggeration, his comments probably contain a good deal of truth:

> Today, one can have no idea of what role the Carnot principle played in our generation. It was the only complete cycle, the standard by which everything was measured (cf. the literature of that time, especially Zeuner). It was the creed of thermodynamics, and to doubt it was blasphemy and heresy worthy of burning at the stake. If one follows the literature further, he will find that it was first the controversy over the diesel engine that shook the position of the Carnot process and brought clarity into the matter. Today, it is easy to be derisive about these developments.[39]

The decade (1880–89) Diesel spent in France as manager of Carl Linde's refrigeration business was the immediate background to the invention of the diesel engine. In his capacity both as businessman and repairman, he learned much about refrigerators and the heating, compressing, and cooling of gases, especially ammonia, which was used in Linde's machines. During the early 1880s, Diesel worked on the invention of a clear ice machine that probably used ammonia; and in 1882 he entered into correspondence with Heinrich Buz, head of the Augsburg Engine Works, concerning the testing of this machine. Such tests never came about, but some ten years later the Augsburg firm would develop the first diesel engine.[40] Sometime in 1885 Diesel abandoned his work on the ice machine because, as he explained in a letter to his parents, his contract with Linde forbade him to exploit any invention for the improvement of refrigeration.[41]

From December 1883 to May 1887, Diesel produced numerous calculations and sketches for three different versions of an ammonia vapor engine (*Ammoniakmotor*, or $NH_3$ Motor). On the basis of the third sketch, he began to construct and test an engine in May 1887. Tests ran until April 1889, at which time the project was terminated because of unsatisfactory results.[42]

At the same time Diesel was abandoning his ammonia engine, he was, however, already formulating the theory of the diesel engine, which was principally completed by the time he left Paris in 1890. "I did not, however, find the time in Berlin to draw out the quintessence of my extensive labors and to put together the results of my research."[43] By early 1892 Diesel had found the time and had written a sixty-four-page manuscript entitled "Theorie und Construction eines rationellen Wärmemotors." Although this manuscript, plus additions, was published the following year,[44] a five-page conclusion entitled "*Schlussbemerkungen*"—"Concluding Remarks" (see appendix 2)—was omitted. Why this occurred cannot be precisely determined. The conclusion was written at the same time Diesel was preparing his first patent draft and the necessity of proving originality must have been prominent in his mind. By early 1893 he had acquired a patent and the need to show originality was less urgent. Also, the book was meant at least in part to be a promotional work that would produce additional financial backing. Diesel may have felt the historical material would be of less interest to potential backers.[45] In any event, the "Concluding Remarks," along with his manuscripts on the ammonia engine, give a picture of how he moved from the ammonia engine to the diesel engine.

Diesel indicated he first conceived the idea of simply substituting ammonia for water in a conventional steam engine.[46] Other inventors at the time were also working with ammonia as a substitute for steam because of ammonia's low boiling point. Therefore, a given expenditure of heat would provide higher pressures than with steam.[47] Even at this early stage, Diesel was concerned with coming as close as as possible to the Carnot cycle. He realized his engine would not operate with isothermal expansion (*"Wärme-zufuhr bei const. Temperatur"*), but thought the greater temperature differential would provide better results than in a steam engine using water. At this point, he seemed to be concentrating on increasing the thermal efficiency by expansion of the piston as far as possible to exploit a large temperature differential.[48]

Diesel was soon experimenting with superheated steam, an idea he may have derived from contemporary attempts to construct functional steam engines utilizing this approach.[49] He hit upon the concept of allowing the steam that escaped the cylinder to be absorbed by glycerin. The glycerin would be heated and then circulated through a coil or mantel and in turn be used to heat more ammonia. Once the glycerin was heated, the engine would be capable of running for ten to twelve hours before all the stored ammonia had been heated and absorbed in the glycerin. The ammonia and glycerin would then have to be separated, a process that would take about two hours. After the two substances were separated, the engine was ready to use again. Diesel projected his engine at about 2/25 hp (6 kilogrammeters per second).

Diesel saw several advantages of his engine. After its initial heating, it would be continually ready to run without warm-up. Superheated ammonia would provide double the work of ordinary steam. The high pressures that resulted would allow for a smaller engine.[50] The glycerin was a good medium of absorption, possessed a high boiling point, and would serve the additional function of acting as a lubricating material.[51]

Diesel sketched out three models of what he called his "intermittent absorption engine." Apparently, at first he thought it could perform several tasks, including heating and ice manufacture, but he soon concluded that such a project was too complicated. Therefore, "the engine should serve one purpose: the production of power."[52] At the end of the sketches for his second model, he undertook a study of high-speed steam engines and concluded that a single-acting, single-cylinder engine was sufficient for small powers. For larger powers, a compound engine of several cylinders was necessary—an idea that would reappear later in his diesel engine theory.[53]

Diesel began to construct an actual engine between July and October 1887. In June 1888 he began to test it. During the nearly one year of tests, he reconstructed it several times. Problems were encountered with noise and shocks, leaks, unsatisfactory heat exchange, blockage of the flow of glycerin, and too little power production. Although Diesel was able to overcome some of these problems, his final note on April 12, 1889 reads: "One sees in general a significant improvement of the whole mechanism, but one cannot yet be satisfied."[54]

The process by which Diesel moved from his ammonia engine to the theory of the diesel engine is best described in the unpublished "Concluding Remarks" to his 1892 manuscript. Because this document is vital to understanding Diesel's thought process, quoting major sections of it is worthwhile.[55]

Diesel began by describing his original plan of constructing a small ammonia vapor engine that would be continually ready to operate and would run for ten to twelve hours after a short initial heating:

> Therefore, I chose fluid ammonia as a working medium and was led to an investigation of ammonia vapor. I did this from a practical and theoretical point of view, made very thorough tests with ammonia vapor, its absorption in various fluids, etc., and constructed an actual ammonia engine. Theory and practice had already led me to the superheating of steam. My experiences with the constructed engine made me acquainted in a surprising way with the advantages of superheating. I drew up a complete theory of a steam engine with highly superheated ammonia vapor and discovered mathematically significant advantages over the present steam engine. This engine distinguished itself in comparison to present day steam engines through its extraordinary smallness, which meant that *not only high temperatures but also very high pressures had to be used for the most advantageous execution of the process.* In order not to go

astray, I also calculated for engines that would use highly super-heated water steam. Here too I saw the necessity of using high pres-sures, *since only through a great pressure differential during expan-sion can a large temperature difference be effectively utilized.*[56]

This represents Diesel's first important idea. His experience with superheated ammonia vapor in a small engine led him to the idea of utilizing high pressures in order to effectively exploit the heat of the vapor "since without it [a great pressure differential] the vapor will remain superheated at the end of the expansion and a part of its added heat will be discharged unused."[57] His calcula-tions showed further that pressures of fifty to sixty atmospheres (atms) were necessary. Under these conditions, however, the han-dling of ammonia and other vapors became too complicated:[58]

It became apparent that the entire scientific material that we pos-sess concerning gases was not sufficient for pursuing the problem. I hypothetically calculated the Regnault steam tables to very high temperatures, and it turned out that for our situation the critical point was exceeded, so that one could no longer differentiate be-tween the condition of a fluid and a gas. This led me to the idea of regarding steam as a gas, if only to approach it more mathematically. Moreover, I discovered that practically no difference exists between steam and gas, *that I therefore could use gas or air. However, I retained the high pressures and temperatures from my earlier investigations.*[59]

This is Diesel's second significant idea. The difficulties in work-ing with ammonia under high pressures and his own calculations led him to substitute air for ammonia as a working medium, while retaining the idea of high pressures and temperatures. As he said in his 1897 Kassel speech:

The effort to replace ammonia vapor with something cheaper and easier to handle led to the use of air. Theoretical investigation led to the same results: *here also a pressure differential, which the-*

*oretically can be described completely, is necessary to utilize a large temperature difference. The two conditions are inseparable.*[60]

When he had worked with ammonia and glycerin, Diesel had put both fluids in separate containers and heated and cooled them from external sources. At first, he did the same with air,

until finally I got the idea of using air not only as a working medium but also as a chemical medium for combustion . . . *I came, therefore, by way of a long detour to an idea that has been used for a long time in gas and hot air engines: combustion in the cylinder itself.*[61] With high temperatures, however, it is not possible to use the normal combustion process advantageously. *Therefore, I got the idea of undertaking combustion in the highly compressed air itself. The pursuit of this idea led to the theory of combustion presented here and to the proposed engine.*[62]

These quotes reveal Diesel's third and fourth ideas: combustion inside the cylinder, using air not only as a working medium but also as an agent of combustion; and the idea of compression ignition, that is, igniting the fuel in the cylinder by means of the high pressures and temperatures of the air. External ignition sources could, therefore, be dispensed with.

At the end of his 1892 manuscript, Diesel summarized his move from ammonia to diesel engine and expressed his ultimate goal:

From the preceding information, one can understand why the final result had to wait for more than a decade. First came a completely new study of superheated water and ammonia vapor with the calculation of hundreds of numerical examples and the preparation of a large number of new tables. Then came a series of tests on ammonia absorption which required the setting up of a whole laboratory and an expenditure of nearly three years. There followed a theory to accompany these tests, then the construction of a real ammonia engine using various types of control mechanisms. Then came an investigation of saturated steam under pressures of several hundred atmospheres and very high temperatures. Next came the statement of a theory of combustion, and finally the design of the proposed engine. A theoretical investigation lasting a number of years led to this whole procedure, which was directed toward a more practical goal, namely, expressing the characteristics of gases, steam, and fluids through unified laws, *since only in this way did I hope to approach the goal of more rationally utilizing the heat of our fuel.*[63]

Before analyzing the complete theory of combustion that Diesel elaborated, it is appropriate to question just how original he was in creating its foundation. Certainly, the elements he purports to have worked out over such a long period—use of air as working medium, high pressures and temperatures in a working cylinder, and compression ignition—were all well known by the late 1880s. Carnot had suggested the use of air, and, by Diesel's time, numerous engines, including the otto engine, were operating with that medium. Compression ignition had been suggested by Carnot, and engineers were aware that higher pressures in an engine cylinder would increase thermal efficiency.

Involved here is a problem that underlies much of Diesel's inventive work. He would continually protest his originality and his ignorance of theoretical and practical developments that seemed to predate his own theory. It has already been pointed out that he may have felt the need to appear original because he was working on his first patent draft at the same time he was finishing his 1892 manuscript. He may, therefore, have wanted to portray his invention in as original a light as possible, especially for patent purposes. Yet, the shortest distance in invention is not always a straight line; often, insights are arrived at after complicated, roundabout thinking and experiment. Diesel may well have been telling the truth when he said that only by a long detour had he arrived at generally well-known procedures. No invention is immaculately conceived, but consists of combining existing ideas.[64] Nonetheless, Diesel's reluctance to admit debts to others would cause him much trouble later on, especially concerning his originality.

One authority that Diesel was willing to acknowledge was Gustav Zeuner. It must have been in the late 1880s, when Diesel was working out his new theory of combustion, that he came under the influence of Zeuner's thermodynamics textbook. It stressed the importance of trying to approximate the Carnot cycle, thus reinforcing an idea that went back to Diesel's student days. Diesel was also struck by several other assertions in the volume. Speaking of the problem of determining the mechanical equivalent of a fuel's heat, Zeuner says: "Here unfortunately one comes across a problem which the present state of thermodynamics does not have the means to solve." Concerning combustion, Zeuner says: "With the present state of thermodynamics there seems little prospect of quickly solving the problem of how combustion should occur, that is, which shape the combustion curve must take to attain maximum work."[65]

Diesel may have been stimulated both by his own experiences as well as by Zeuner's book to reach his major inventive insight. It was all very well to have combustion in highly compressed air, but the rise in temperature during combustion would mean much lost heat energy, the same problem that could be seen in the otto engine. What if one could, in fact, approximate the Carnot cycle in a heat engine, that is, achieve combustion and expansion at a constant temperature? Then, almost all heat produced would be converted to useful work. As Diesel said in his "Concluding Remarks":

> The proposed new engine is independent of whether the specific heat of the gases is variable and of other questions which are raised by theoreticians as preconditions for the investigation of combustion. *Fundamentally, the Carnot theorem is the result of my investigations, and it is completely independent of the [physical] constants, indeed, of any of the substances employed.*[66]

Diesel was saying here that his theories would lead to an engine that employed the Carnot cycle. Further, he thought he could bypass some of the problems associated with the combustion process.

He was, therefore, forging ahead into new, untried waters. In fact, problems with proper combustion would plague the diesel engine for some years.

The question of how to approximate the Carnot cycle in practice, however, was more difficult. Diesel thought that other heat engines, such as gas and hot-air engines, all had a false working principle: the amount of air and fuel mixed was almost equal. Combustion thus caused a major temperature increase, and much of the heat was lost through the cylinder walls or through exhaust gases. In other words, the combustion process was uncontrolled and not rational—combustion temperature was determined during the process of combustion itself.[67]

Suppose in contrast to these principles, that considerably more air was drawn into the cylinder than was needed for combustion, in fact, about 100 kilograms (kg) of air for every one of fuel, or nine times the amount of air needed for combustion. Further, the air would be greatly compressed so that combustion temperature would be reached by mechanical compression alone and not be exceeded during the combustion process. The temperatures and pressures attained would be so high that ignition temperature would be substantially exceeded and compression ignition would be the result.[68] At the point of maximum compression (top dead center), a small amount of carefully measured fuel—Diesel was thinking at this point primarily of powdered coal—would be gradually introduced into the highly compressed air and would ignite and burn. The tendency for the temperature to rise would now be exactly compensated for by the expansion of the air, which would absorb the generated heat, thereby producing combustion at a constant temperature, that is, isothermal combustion (or expansion).[69] After the addition of fuel is discontinued, expansion continues adiabatically (that is, with a drop in temperature of the working medium), as close as possible to atmospheric temperature. The cylinder walls would not need to be cooled; instead, they would be insulated to guard against heat loss. For maximum results, compression should proceed at first isothermally, which can be achieved through the injection of water into the cylinder, and then adiabatically (that is, with a rise in temperature of the working medium).

The result of this process is an approximation of the Carnot cycle: isothermal compression is followed by adiabatic compression. Isothermal expansion is then followed by adiabatic expansion. Diesel originally proposed maximum pressures of 250 atm and a high temperature of 800°C or 1,073° Kelvin. The lowest temperature was to be 20°C or 293° Kelvin. Because, according to the theory first worked out by Carnot and refined by later experts, the thermal efficiency of any engine is the maximum temperature minus the minimum temperature divided by the maximum temperature (all temperatures being given on an absolute scale), Diesel's engine was to have an efficiency of 72.7 percent.[70]

As Lynwood Bryant has suggested, Diesel's central bright idea was to figure out a way of achieving isothermal combustion in a cylinder through the use of far more air than was necessary for the combustion process.[71] "The result is fundamentally a hot air engine which almost exactly carries through the Carnot cycle."[72] Diesel was especially proud of his theory; everything could be exactly calculated mathematically, thus producing a rational combustion process that would convert the maximum amount of heat into work.

As early as his February 1892 patent draft, Diesel was examining alternative combustion procedures. For instance, he mentioned the possibility of purely adiabatic compression: "Also, combustion can be conducted somewhat differently than purely isothermally."[73] Already, he seemed to realize that his process might not occur completely at constant temperature and that some temperature increase might occur during combustion. He still held, however, that isothermal expansion was the ideal to be realized as closely as possible.

When Diesel came to consider the design of an actual engine, he repeated the idea that a single-cylinder engine would be adequate for smaller powers. It would operate on the four-stroke cycle; no ignition device would be necessary. Compression would be accomplished through a rapidly moving plunger piston, "quite similar to the familiar atmospheric lighter."[74] At this point, Diesel uses the pneumatic lighter, known since his school days, to illustrate compression ignition.

In his 1892 manuscript, as well as his book, *Theorie und Konstruktion*, Diesel proposed fabricating a 100-hp three-cylinder compound engine, undoubtedly because he thought the required pressures and expansion would necessitate an impossibly large cylinder.[75] Two smaller high-pressure cylinders would operate 180° out of phase with a larger, double-acting low-pressure cylinder. Air would be compressed a few atmospheres in the low-pressure cylinder, then channeled into the high-pressure cylinders, which operated on the four-stroke cycle, alternately compressing and expanding. The air would be compressed to maximum temperature, and then fuel would be added, followed by combustion and expansion. Expansion would continue in the low-pressure cylinder. It would draw in air from below and eject air from the top. Diesel's idea of fuel addition was remarkably simple. Fuel in the form of powdered coal would be stored in conical hoppers above the smaller cylinders. It would then fall into the cylinders through a rotating disc mechanism. "The process is rather similar to the fall of sand into an hourglass."[76]

Diesel once again argued for the originality of his theory. In his "Concluding Remarks," he said: "The preceding investigations and engine to be constructed on their basis certainly did not spring from existing opinions." Toward the end of the manuscript, he remarked: "I never in my life studied a theory of hot air and gas engines. I did this only subsequently to control my investigations. I found, however, the same negative results which are expressed in the previous quotes from Zeuner and Richard."[77]

How honest was Diesel in these assertions? Is Kurt Schnauffer, in fact, correct when he claims: "The originality of Diesel's invention is undoubtedly that he arrived at his new engine, completely independent of any kind of precedents, exclusively through correct combinations which were based on thermodynamic calculations."[78]

In 1887, while Diesel was still in Paris, Otto Köhler, a teacher at the mechanical engineering school in Cologne, had his *Theorie der Gasmotoren* published.[79] In this work, he proposed a three-cylinder engine that is extremely close to what Diesel proposed several years later. One cylinder compressed air, another gas; then the two were admitted to the working cylinder and "ignited in some way" (*"in irgend einer Weise entzündet"*). The heated air would then drive the piston forward. Compression would be both isothermal, by means of water injected into the cylinder, and adiabatic. Expansion would be isothermal, because of the use of an excess amount of air over that needed for combustion, and then adiabatic. Cooling of the working cylinder would not be necessary. Such an engine would approximate a Carnot engine.[80]

Obviously, the principles of Diesel's engine are similar to those of Köhler's, though the two three-cylinder engines do not work exactly the same; nor does Köhler mention compression ignition, an important part of the diesel process, even if Diesel himself did not regard it as *the* defining element. Further, Diesel had mentioned the idea of a compound engine in connection with his ammonia engine, and, in his brief sketch of a solar engine, he had suggested cooling the working medium by the injection of water.

Köhler's ideas were also picked up and described by Franz Grashof in his 1890 *Theorie der Kraftmaschinen.*[81] Lynwood Bryant has argued that Diesel could hardly have overlooked the book by Grashof, an eminent authority in the field.[82] It is obviously possible that Diesel picked up some of his key ideas from Köhler's book or Grashof's exposition of it. Diesel, however, said later that, during the time he was finishing his 1892 manuscript and writing his patent drafts, he was unaware of Köhler's book and other literature on the subject.[83]

This contention seems to be substantiated by Paul Meyer, one of his early coworkers and author of a perceptive manuscript on the history of the diesel engine. Although often highly critical of Diesel,

Meyer argued that he was probably not familiar with Köhler's ideas in 1892 because they would have caused major problems for him in terms of his patent application and publication of his *Theorie und Konstruktion.* Meyer further quotes a letter from Moritz Schroeter of October 1893, when Köhler's ideas and his criticisms of Diesel had become public, expressing astonishment after reading Köhler's book that his and Diesel's ideas were so close.[84]

If Schroeter, one of Diesel's former teachers and a highly respected professor of mechanical engineering, had overlooked Köhler before 1893, it is also possible that Diesel could have done the same. Whatever Diesel's knowledge of Köhler, he was to say in 1897, not without some truth, that, though some of his ideas had been presented in earlier literature, "there was never any description of a process in its totality which brought together the requirements of theory with the needs of practice, and therefore could have produced a new engine type."[85]

Diesel's engine must have owed something to earlier experimentation with oil engines that involved high compression and various ignition devices. Once again, Diesel tried to maintain his innocence of such activities. For example, in an undated manuscript from the period of the development of his engine, 1893–97, he presents sixteen pages of notes on Gustave Richard's *Les moteurs à petrol depuis 1889.* All recent engines are mentioned, and Diesel adds in a marginal note: "The engine for which I am striving is actually the combination of the Akroyd vaporization chamber and the Brayton injection device. I emphasize, however, that I learned of these constructions in November 1893, after the Augsburg test engine was long in operation."[86]

Diesel is referring here to an engine developed by the American George Brayton during the 1870s that for the first time used pressurized air to inject and vaporize the liquid fuel. Friedrich Sass believes Diesel picked up the idea of pressurized fuel injection by studying Brayton's engine.[87] The English inventor Herbert Akroyd Stuart developed in the early 1890s an engine in which oil was injected into a vaporizing chamber and then mixed with air. Ignition was generated by the hot surface of the vaporizing chamber.[88] His engine is sometimes referred to as a hotbulb or "semi-diesel." Several other engines developed by such men as Julius Söhnlein and Emil Capitaine have also been mentioned as precedents. These engines involved vaporization of fuel, high compression, and ignition by hot-bulb or through the surface of the vaporization chamber. As Sass says, however, Diesel was working with higher pressures—his original theory called for 250 atms, which was eventually reduced to 44—nor did any of these engines operate with true compression ignition. Just how informed Diesel was about these earlier engines is difficult to say. He said in his "Schlussbemerkungen" that "neither at the Paris Exposition of 1889 nor at the Frankfurt Exposition of 1891 was any engine exhibited which deviated from three types: steam engine, gas engine, or hot air engine."[89] Such a statement seems to indicate that he was familiar with these engines by 1889 and seems to contradict other assertions in the same document that he had never studied a theory of gas engines or hot air engines in his life until after working out his theory. Once again, this may be a case of Diesel magnifying his claims to originality for patent purposes. Diesel was correct, however, in the assertion in his Kassel speech of 1897 that he was the first to take a number of ideas, organize them into a complete theory, and from these develop a working engine.

Otto Köhler not only described a theory very similar to Diesel's ideas, but also correctly pointed out the major problems with Diesel's theory. Köhler was apparently made aware of Diesel's ideas in 1893, when a mutual friend, A. Venator, an engineer of the Linde Society, wrote him at Diesel's request asking for an opinion on Diesel's newly published *Theorie und Konstruktion*. Diesel was at that time requesting opinions on his theory from numerous experts in the field. Köhler answered Venator on March 18, 1893, with a negative opinion. His ideas were repeated in a speech to the Cologne VDI on April 10, 1893, which was printed in the *VDI-Zeitschrift* in September 1893.[90] Köhler would make trouble for Diesel later in 1897 by threatening legal action against his patent.

Likely, because of Köhler's criticisms, Diesel modified his theory during the summer and fall of 1893. After drawing attention to the similarity of his 1887 ideas to those of Diesel's, Köhler went on to point out that the engine he himself had proposed in theory would not work in actuality:

These examples lead to the conclusion that the ideal cycle (quite apart from the difficulties in executing it exactly) with air as working medium is not suited for practice. To be sure, a higher thermal efficiency is attainable; however, the initial pressure rises so significantly without $p_i$ [the mean pressure] rising accordingly, that the measurements of the cylinder and crank drive result in disproportions. In addition, the thermal gain is more than used up by the great losses due to friction.[91]

Köhler pointed out that there was a maximum pressure which it was advantageous to utilize in an engine. Exceeding this pressure would cause thermal efficiency to be more than offset by losses in mechanical efficiency, which is brake thermal efficiency divided by indicated thermal efficiency. Köhler thus touched on Diesel's major problem. The latter was too concerned with thermal efficiency; he either ignored or skipped over other efficiencies, such as brake thermal or mechanical. Köhler also maintained that isothermal compression and expansion would be extremely difficult to realize in an actual engine involving high speeds. Further, the high pressures and expansion Diesel required would result in a large cylinder. In fact, all the work gained through expansion would be more than used up in the negative work needed to compress the air— the engine would not even be capable of turning over. Köhler ended by saying much better ways of utilizing fuel were available than through the Carnot cycle.

Köhler made one incorrect assumption: that compression ignition had to be abandoned. Otherwise, his criticisms were on the mark. Although Diesel's theory was in itself correct, it was not suitable for a real engine. To put the main criticism another way: Die-

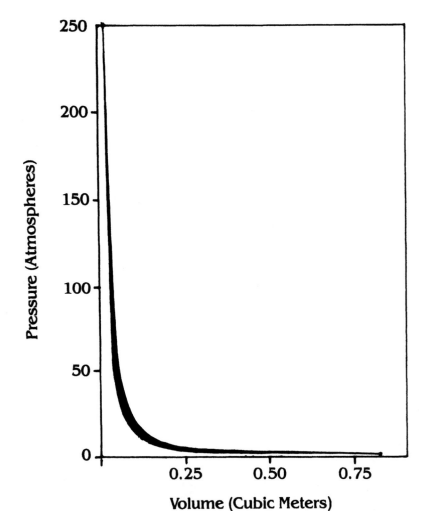

FIGURE 2. PRESSURE-VOLUME DIAGRAM OF DIESEL'S IDEAL CARNOT ENGINE.
Compression of the working medium to a pressure of 250 atmospheres (atms) or 3674 psi
was well beyond the technology of Diesel's time. Further, the very narrow area of net
work between the expansion and compression lines should have indicated to Diesel that
the engine was not capable of developing any useful work. Drawing by Betty Spencer,
adapted from Diesel's *Theory and Construction of a Rational Heat Engine*, the 1894
translation of his theoretical treatise.

sel posited a major excess of air to realize isothermal combustion. He overlooked the cost in negative work of driving all that excess air back to its compressed stage. So little fuel was used in relationship to the quantity of air that not enough power would be generated to run the engine. Indeed, the theoretical indicator diagram that Diesel used to illustrate his theory showed the expansion and compression lines so close together (see figure 2) that it is surprising this did not give him cause for concern. Why Diesel could not see the problem with his theory is unclear. As Kurt Schnauffer says, mechanical efficiency was already a well-known concept by Diesel's time. Some consideration of it would have quickly convinced him he was positing too much air to achieve workable results. Perhaps the explanation lies in the fact that "he was a newcomer in the area of internal combustion engines. He, therefore, had no experience and turned too little to practical men for advice. . . . It was an unsuitable attempt to control combustion in a newly invented engine by a specific law."[92]

Whatever the reasons for Diesel's miscalculations, by 1892 he had completed his theory. He was now ready to apply for a patent and seek industrial backing.

### German Patent Law and
### Diesel's First Patent

On February 26, 1892, Diesel drew up his first sketch of a patent application. The patent itself, "German Imperial Patent No. 67207," was not granted until December 1892. During the year that passed between original sketch and final issue, the patent was subject to significant revisions. Because of the controversy over Diesel's patents and the role they played in the invention, development, and innovation of the diesel engine, a brief history of German patent law up to the time of Diesel's first application is appropriate at this point.

The first imperial German patent law went into effect in 1877; it was a result of German unification and replaced the patent laws of the individual states. The law was passed just before the beginning of the era of protectionism in Germany and came at a time of increased technological complexity, usually requiring large capital investment to back inventive efforts. Such investment increasingly could only be provided by large-scale industry or investors. Through such requirements as high fees, the patent law hindered lone investors from becoming entrepreneurs. Increasingly, firms that provided capital and equipment for inventive efforts demanded that the patents be taken out in their names. Many of them also acquired the right to control the patents or exclusively exploit the inventions of employees.[93] This feature impeded Diesel's inventive efforts with his clear ice machine. As time passed, apparently problems were encountered with the law, particularly the process of judging the validity of patent applications and the invalidation proceedings against granted patents, as well as lack of clarity in the exact definition of what an invention was.[94]

A good illustration of the early problems in German patent law was the successful invalidation proceedings in 1886 against the otto engine patents. The historian of the otto engine, Gustav Goldbeck, points out that the brief period of development of German patent law and lack of experience may have contributed to this decision. Furthermore, it must have been difficult for the judges to follow technical arguments about exactly what went on during combustion in an engine—an area at that time still open to much uncertainty. The judges made their decision on purely formal grounds without relating the otto patents to the historical development of the engine. As advances were made, patents were taken out to cover them, but it could be argued that by then numerous engines were working by these principles and no novelty was present. Even Ernst Körtung, one of Otto's opponents, referred to the arbitrariness in patent cases as "legal decisions by lottery."[95]

To overcome these problems, a revised patent law was drawn up and went into effect in early 1891, approximately a year before Diesel applied for his patent. Major features of the law were: (1) patents would be granted for inventions that permitted an industrial utilization; (2) patents could be granted for new processes or procedures as well as actually existing machines; (3) an invention was not valid if it had been described in print within the last hundred years; (4) the length of the patent was fifteen years; (5) invalidation proceedings could only occur during the first five years; and (6) the fee for a patent was 50 marks, rising 50 marks a year to a total of 700 marks by the fifteenth year.[96]

Despite the new law, difficulties continued to exist. The problem of exactly defining a new invention and its novelty persisted, just as it does today.[97] Even one of Diesel's critics, Alois Riedler, devoted an appendix in his book on the diesel engine to a discussion of patent law. He criticized the capability of patent officials; the preliminary testing procedures; the high fee scale; and the problem of determining printed precedents within a hundred years, regardless of how obscure or whether any attempt had been made to turn the ideal into reality.[98]

Statistics from this period indicate that, though the number of patent applications increased fairly steadily from 1877 to 1890, the number of patents granted each year during the same period remained almost exactly the same. The *VDI-Zeitschrift*, which published these statistics, attributed this situation to patent applicants not searching out previous literature and the patent office not making its preliminary examination procedures clear enough nor its library sufficiently available.[99] Thus, Diesel was applying for a patent during a period in which many patent applications were being rejected, only six years after the otto engine patents had been nullified, and in which much uncertainty existed about what actually happened during combustion in an engine's cylinder.

Diesel's first patent application ran fifteen handwritten pages. Paul Meyer has remarked that the first draft does not give much evidence that he was very experienced in writing patent applications.[100] He spent the first five pages exposing what he regarded as the false working principles of hot-air and gas engines. After expressing the principles of his new theory, he spent only a few pages on actual engine design. He then presented the advantages of the

Neue, rationelle Wärmekraftmaschine
von Rudolf Diesel.
Ingenieur. Berlin.

Es sind heute 3 Arten von Motoren bekannt, welche Luft oder Verbrennungsgase, einzeln oder gemischt, als motorisches Medium verwenden, nämlich:

A. Feuerluftmaschinen.

B. Gasmotoren (bzw. Petroleummotoren).

C. Heißluftmaschinen.

§A. Der Grundgedanke der Feuerluftmaschinen ist, in einem geschlossenen Feuerraum Verbrennungsluft unter Druck einzuführen, und die gebildeten Verbrennungsgase arbeitsverrichtend expandiren zu lassen. In die Luft setzt in Gegenwart von glühendem Brennstoff ist, so kommt gerade die zur Verbrennung theoretisch nothwendige Luftmenge zur Wirkung und die Verbrennungsgase enthalten keine oder nur geringe, zufällige, Überschüsse an unverbrannter Luft. Die theoretische Luftmenge bei reiner Kohle ist rund 11,82 kg. pro 1 kg. Kohle, die Summe der Verbrennungsprodukte also rund 12,82 kg. oder unbedeutend mehr. Da nun die Verbrennung von 1 kg. Kohle rund 7800 calorien entwickelt, und die spec. Wärme der Luft rund 0,25 beträgt, so entstehen bei dieser Art von Verbrennung Temperaturen von weit über 2000° C. Diese zwingen dazu, die Wände des Ofens und des Arbeitscylinders energisch zu kühlen, da sonst die Maschinen unhaltbar sind. In dieser Kühlung liegt der erste bedeutende Verlust an Wärme.

Diesel's first handwritten patent application,
February 1892. (M.A.N. Werkarchiv)

engine, including its role in industrial decentralization, and ended with twelve claims for its originality. The first six are summarized here:[101] (1) avoidance of the mixture of air and fuel, as it occurs in open and closed fireboxes, as well as in explosion motors; (2) compression of the air well beyond the ignition temperature of the fuel, which could occur adiabatically, but most completely by a combination of isothermal and adiabatic compression; (3) choice of the exact amount of air so that the desired temperature will not be exceeded during combustion; (4) choice of an exact amount of fuel so that the combustion process takes place as isothermally as possible; great deviation from isothermal combustion is possible without rejecting the essentials of the theory; (5) adiabatic expansion of gases after combustion; and (6) complete avoidance of cooling the cylinder walls.

Kurt Schnauffer sees the first two claims as the most important because they are essentially realized in the diesel engine today.[102] Yet, in terms of Diesel's 1892 theory, all the claims fit together in one theory of rational combustion. It is not quite fair to chastise the patent officials, as Schnauffer does, for not recognizing this and for not protecting the first two claims at the expense of the others.

To handle his business, Diesel contacted the well-known Berlin patent attorneys F. C. and E. Glaser. His first patent application was pronounced unsatisfactory and returned by the Patent Office on March 15, 1892.[103] Rewritten applications of April 6, May 12, and June 22 were also rejected. The decisions by the Patent Office indicate that the applications did not show how the claims could be realized in an actual working engine. Drawings were requested that would show how Diesel's principles could be put into practice. In its decision of June 4, the Patent Office claimed the ideas of "the introduction of fuel into compressed air" and "an uncooled powdered coal engine with compression ignition [are already] known." Still not satisfied with Diesel's revised applications, the Patent Office on July 7 recommended wording that reduced the patent claims to two. In a renewed application of July 23, Diesel used almost the exact wording suggested by the Patent Office. On September 3 the new patent draft was accepted and announced publicly; a four-week period was allowed for objections. Because none were made, the patent was granted on December 23, 1892. It was to run from February 28, 1892, the date of first application, and was officially published on February 28, 1893. Its expiration date was February 27, 1907.

The first claim in the final version read:

> Working process for an internal combustion engine characterized by the fact that in a cylinder pure air or another neutral gas (or steam) mixed with air is so highly compressed by the piston that the resulting temperature is far above the ignition temperature of the fuel, . . . whereupon addition of fuel beginning at top dead center occurs so gradually that combustion takes place without essential pressure or temperature increase because of the outward moving piston and the expansion of compressed air that this brings about. . . . after the conclusion of fuel admission, the further expansion of the quantity of gas in the working cylinder takes place.

The second claim described Diesel's three-cylinder compound engine, in which his combustion process was to be carried out.[104]

In the sense that the wording of the first claim links the compression of air and then the introduction of fuel at maximum compression directly with isothermal combustion, the patent becomes more open to attack than if the claims were separated, as in the first draft. In this way, the Patent Office unwittingly weakened the patent. Yet, Diesel's theory had indeed linked all these elements. The final patent version, therefore, may more accurately summarize his ideas. Unfortunately, he accompanied his inventive efforts with a patent that protected a theory of combustion which was not suited for a working engine.

After his experiences fending off legal challenges to his patents, Diesel reflected on the patent law in his *Die Entstehung des Dieselmotors*:

> Our patent law knows in general only the protection of ideas, not the protection of inventions. For that reason, our patent law can

# PATENT-URKUNDE

№ 67207

AUF GRUND DER ANGEHEFTETEN BESCHREIBUNG UND ZEICHNUNG IST
DURCH BESCHLUSS DES KAISERLICHEN PATENTAMTES

*an Rudolf Diesel, Ingenieur,
in Berlin*

EIN PATENT ERTHEILT WORDEN.

GEGENSTAND DES PATENTES IST:

*Arbeitsverfahren und Ausführungsart für
Verbrennungskraftmaschinen.*

ANFANG DES PATENTES: *28. Februar 1892.*

DIE RECHTE UND PFLICHTEN DES PATENTINHABERS SIND DURCH DAS PATENTGESETZ
VOM 7. APRIL 1891 (REICHS-GESETZBLATT FÜR 1891 SEITE 79) BESTIMMT.

ZU URKUND DER ERTHEILUNG DES PATENTES IST DIESE AUSFERTIGUNG
ERFOLGT.

*Berlin, den 23. Februar 1893.*

## KAISERLICHES PATENTAMT.

Beglaubigt durch *Frank*

Bureau-Vorsteher des Kaiserlichen Patentamtes.

Diesel's Patent Number 67207, February 1893.
(M.A.N. Werkarchiv)

destroy the most worthwhile invention, when it is simply shown that the idea was already moldering away in some forgotten writing. . . . a patent is not a scientific treatise, which one can carefully examine under a magnifying glass. The patent text is drawn up on purely practical, technical, or tactical grounds that have nothing in common with science. Often it arises out of a compromise with the examiners, which is quite different from what it ought to be in strictly scientific terms. To want to use and to criticize patent texts as a test of scientific ideas is ignorant foolishness.[105]

Diesel's case demonstrates some of the problems involved in the patent law. The identification of what is new often was (and is) a perplexing problem, especially when anything in the printed literature for the past hundred years, no matter how obscure, could invalidate it. Some of Diesel's ideas, such as compression ignition, had already been anticipated, though he was the first to realize them in a working engine. Furthermore, patenting a new process, without the corresponding working engine in existence, presented difficulties. In Diesel's case, many unknowns existed, the move from theory to practice was no simple matter, and the final working engine did not correspond with his theory as propounded in his patent claims.

In late 1893 Diesel applied for a second patent, which also did not adequately protect his engine. This second patent will be discussed at the end of this chapter. Why did not Diesel in 1897 take out a new patent covering the actually developed engine instead of trying to defend his first patent? As Paul Meyer has pointed out, new patent claims would have been so close to what was already contained in Patent Nr. 67207 that their novelty would have been difficult to prove. In addition, it would have been hard to formulate all the new mechanical details of construction into appropriate claims that would have satisfied the patent law's demands for novelty.[106] Also, Diesel was close to the point where his first patent could no longer be legally challenged—February 1898. If his patent survived until then, there would be no point in trying to gain a new one. Despite its problems, the Diesel main patent is extremely important because it is the basis on which he was able to sign contracts with industry in order to develop his new engine.[107] The attacks on Diesel's patent will be discussed in chapter 5.

### Initial Responses to Diesel's Theory

During March and April 1892, Diesel was not only applying for his first patent, but he was also attempting to win industrial support. Chapter 4 will examine in more detail his negotiations with Heinrich Buz's Augsburg Engine Works, which lasted from March 7 to April 20 and resulted in Augsburg's agreeing to test the engine under very circumscribed conditions.

Even before his first patent draft, Diesel had sent his manuscript to his employer and former teacher, Carl Linde. Naturally, Diesel sought the comments and help of his employer, who was also an expert in the fields of thermodynamics and machine design. As shall be discussed in chapter 4, Linde's refrigeration company was the first model Diesel attempted to emulate in seeking industrial support. Diesel's letter to Linde, which accompanied his manuscript, is dated February 11, 1892, and is interesting because it was the first communication of his ideas to an outside source. He said:

> . . . it gives me great pleasure to tell you that I have discovered an engine which theoretically uses only one-tenth of the coal that our best contemporary steam engines use. . . . This result is not a conjecture or hope. It can be mathematically proven to the point where no doubt can exist that it is attainable. Practically, the construction of the engine offers no difficulties worth mentioning, since it requires nothing fundamentally new technologically. Even if only a thermal efficiency of 60–70 percent of the theory is realized, the machine will use but a sixth or seventh part of fuel now consumed by the best engines.

Diesel then asked Linde to represent his interests, especially with potential industrial backers, and suggested a profit-sharing scheme. Conversations with Linde in February revealed that Linde would not materially back Diesel's project. On March 20 Linde wrote Diesel concerning the theory in his manuscript. The letter was ambivalent, for it alternated between praise and criticism of the inventor's ideas. In essence, Linde praised Diesel for being theoretically correct, but felt the practical difficulties in building such an engine were formidable and that at best he could obtain only about a third of the proposed thermal efficiency. Although Linde offered to help the cause and suggested Diesel was especially well prepared for the task, he also indicated that Diesel had to leave the Linde company as soon as he started devoting himself to the task of building an engine. In all his comments, however, Linde failed to recognize the basic problem with Diesel's theory: the attempt to realize isothermal combustion in a working engine through the use of a large air-to-fuel ratio in the cylinder.[108]

Linde asked for and received permission to send the manuscript to Professor Moritz Schroeter, of the Munich Technische Hochschule, also one of Diesel's former teachers. Schroeter's reply of March 29 stated that, though he too thought the theoretical basis for Diesel's work was correct, the practical difficulties, especially those of ignition, high pressure, and temperatures, could only be solved by long and costly experimentation.[109] Like Linde, Schroeter could see the practical difficulties, but not the problem with isothermal combustion in Diesel's argument.

This first, short, rather noncommittal response would soon be followed by much more positive reactions on Schroeter's part, until he became one of the chief supporters and propagandists for Diesel's engine. Schroeter was one of the leaders of the German movement to replace the steam engine with a more thermally efficient engine.[110] This fact may have swayed him in Diesel's favor. On April 13 Diesel wrote him a letter, a copy of which was sent to the Augsburg Engine Works in hopes of obtaining its support. In this letter, Diesel emphasized that his manuscript dealt only with a theoretical ideal solution and that, for practical purposes, the compression in the cylinder could be reduced from 250 to 44 atms without that much loss of thermal efficiency, and that high temperatures would only exist momentarily at highest compression. [111]

Schroeter answered on May 2, 1892, now much more positive about Diesel's new figures and emphasizing again that he never doubted the theory, only the practical difficulties. However, in the course of the letter, Schroeter, apparently without realizing the total implications of his comments, pointed out that, in Diesel's drawing of a theoretical indicator diagram, the compression and expansion lines were so close together that the area of work diagrammed by the lines was precariously small. Schroeter's logical conclusion should have been that the Carnot cycle was not capable of developing significant work, but apparently he still accepted the idea that the cycle was a valid goal. The question was: how could one enlarge the diagrammed area of work?[112] As Schnauffer points out, both Schroeter and Diesel seemed at this point to be approaching the problem graphically, by working with theoretical indicator diagrams.[113]

Diesel also corresponded during April 1892 with Gustave Richard in Paris and with Eugen Langen. The Richard correspondence is extant in the Deutsches Museum, but is now illegible. It is only known that Diesel wanted to use Richard's recommendation to approach Langen. Langen replied on April 9 that, though he too thought Diesel was theoretically correct, his ideas contained little that was new. Langen thought Diesel would probably not obtain a German patent, but indicated he might be willing to work with him if he did.[114]

By the summer of 1892, Diesel had made significant gains. Experts such as Linde and Schroeter had agreed with his theory but pointed out practical difficulties, which caused him to lower his estimates of temperature and pressures and to speak of realizing a divergent process, while still keeping the Carnot cycle as the ideal to be striven for. Apparently he was willing to compromise with the ideal in an effort to win industrial backing. Although some industrialists spoke critically of his ideas, he had won preliminary support for his project from the Augsburg Engine Works—which support, however, was not yet spelled out in a contract. His patent was reworked to the point where it would soon be accepted. In Schroeter's letter of May 2, 1892, the first inkling of problems with attempts to approximate the Carnot cycle had surfaced. What is interesting from today's perspective is that no one—neither academicians nor industrialists—had questioned Diesel's attempt to achieve isothermal combustion through an excess amount of air in the cylinder. Can one then be surprised if Diesel did not at first see this flaw either?

### Diesel's Book Theorie und Konstruktion

By the fall of 1892, Diesel's patent application had been accepted, and he had decided to publish his manuscript in book form to attract further support for his inventive efforts. The manuscript was expanded by several chapters to deal with modifications of the ideal cycle because he knew by then that industrialists would be frightened off by his original calculations. The modifications he mentioned dealt primarily with dropping isothermal compression and adopting purely adiabatic compression, which would allow much lower compression values—from 250 atms down to 90. Diesel still believed, however, in isothermal combustion.

Diesel sent his manuscript to Springer Verlag in Berlin on October 2, 1892. In an accompanying letter, he claimed that his engine was patented in all industrialized countries and was currently being constructed in Augsburg. He then quoted favorable opinions from letters of Linde and Schroeter and claimed to have similar comments from industrialists, even in America. "This engine," he said, "is destined to bring about a complete revolution in engine construction and to replace existing engines." As Paul Meyer has pointed out, Diesel was twisting the facts to serve his purposes. His engine was not yet patented in all industrialized countries, he had only quoted favorable opinions while suppressing the unfavorable ones, and he apparently had made no contact with anyone in

America. Yet, the letter had its intended effect. The very next day, October 3, Springer replied that they were interested in publishing the book and invited Diesel to discuss the project. Only two days later, they sent a draft contract to Diesel, who signed it and sent it back on October 11.[115] In a letter of October 12, Diesel informed Augsburg of his publication plans and emphasized the idea of generating publicity for his ideas in order to open the way to commercial exploitation.[116]

Diesel's *Theorie und Konstruktion* appeared in January 1893. Of 1,000 copies printed, 200 were still unsold in 1897. Diesel himself bought back from the publisher a number of copies after he discovered his errors. The edition was sold out in March 1898. Springer approached Diesel about a second edition, but was told by one of the officials of Diesel's design bureau that it would have to wait because of the burdens of work. This was undoubtedly true inasmuch as Diesel was making preparations to exhibit his engine at the Second Power Machine Exhibition, in Munich. It is also the case, however, that the book would have needed to be completely rewritten because the developed engine deviated considerably from the theory contained in it. Such a rewriting would have played into the hands of those attacking Diesel's patent, and so the book was never republished.[117]

An English edition was published in 1894, which listed Bryan Donkin (1835–1902) as translator. In fact, the actual translation of the book was undertaken not by him, but by his wife, who knew German. Diesel worked with her in Augsburg in July 1893 on a word-for-word translation.[118]

Donkin's grandfather, after whom he had been named, had been a well-known engineer and founder of a firm, which Donkin took over in 1889. He had become an expert in the construction and testing of steam boilers and heat engines, and in 1894 he also published a textbook on heat engines that had gone through five editions by 1911. The available evidence does not indicate how Diesel became acquainted with Donkin, but the latter became an early supporter of the diesel engine cause in England.[119]

Copies of Diesel's book were sent to a variety of academicians, industrialists, and bankers. Included were Professors Zeuner, Riedler, Schroeter, Reuleaux, Gutermuth, and Helmholtz of the Physikalisch-Technische Reichsanstalt (PTR); Krupp, Siemens and Halske, and Wilhelm Oechelhäuser, of the Continental Gas Company; Langen; Sulzer Brothers, in Switzerland; and the Deutsche Bank and Disconto-Gesellschaft, in Berlin. In letters to the academicians, Diesel stressed that his engine would convert almost 73 percent of the heat of fuel into work and that it would be much smaller and simpler than the steam engine. In letters to industrialists and bankers, he played down construction difficulties and emphasized that his engine, rather than being a new gas type, was designed to replace steam and other internal combustion engines. In all his letters, he once again selectively quoted from previous correspondence with Linde, Schroeter, and Zeuner to make it appear as though they had made only positive judgments of his work.[120]

With a few exceptions, the replies from the professors were quite positive. For example, Reuleaux answered on March 22, 1893: "The theoretical working through seems to be flawless and highly convincing." Concerning practical problems in constructing the engine, he said, "I see no forbidding obstacles ahead." Schroeter replied that he hoped Diesel, like Linde before him, would soon appear with a marketable product that would dethrone the steam engine at the end of the very century that had enthroned it. Zeuner proclaimed both of Diesel's fundamental ideas correct: the heating of the air through compression to combustion temperature; and the use of large quantities of air to achieve isothermal combustion.[121] Already, in February and March 1893, Schroeter and Professor Max Gutermuth, of Aachen, respectively, had articles published that were panegyrics of Diesel's proposed engine.[122]

In his article, Schroeter welcomed Diesel's daring new attempt "to come considerably nearer to the ideal of the Carnot cycle than has been heretofore possible." He recognized that it would only be gradually possible to approximate the cycle, but "we have here the gratifying example of theory outstripping practice and indicating exactly the right methods for achieving the goal. Machine technology must now show how it proposes to realize the theoretical ideals."[123]

What is striking is that Diesel had won over four of the foremost authorities on thermodynamics and engine design in Germany. Of course, how thoroughly they had read his book or worked through his calculations cannot be determined. Perhaps Paul Meyer was correct when he said of Zeuner and Schroeter that they "only examined the diesel engine—or better said rational heat engine— from the standpoint of thermodynamics, without more closely investigating the technical feasibility and practicability of the process proposed by Diesel."[124]

The industrial reaction to Diesel's theory will be examined in more detail in the next chapter. It can be said here that it was far less positive than the response of the professors. Langen, for instance, remained as negative as the year before.[125] Only Krupp was positive. Industry hesitations, however, hinged on practical, not theoretical, considerations.

A number of people did criticize Diesel's theory. Otto Köhler pointed out its chief problems. Alois Riedler, as well as an engineer from Sulzer Brothers, in Switzerland, also thought that the work needed to compress the air in the cylinder and the accompanying loss of heat would cancel the positive work. Gustave Richard, in Paris, apparently also reacted negatively, judging from Diesel's comments in subsequent letters to him.[126]

Partly because of the favorable reactions of many academicians and despite the attacks of others, Diesel signed contracts with Augsburg in February and with Krupp in April 1893. The first series of tests were held from July 17 to August 22 of that year in Augsburg.

### Diesel Arrives at a Workable Theory

Already, in May and June 1893, however, before the beginning of engine testing, Diesel had undertaken calculations that finally produced the outlines of the correct combustion process. The impetus for these calculations appears to be the comments of Schroeter in 1892 concerning the narrow indicator diagram for Diesel's ideal cycle and the criticisms of Köhler. One of Diesel's concerns since the Schroeter letter was how to increase the area diagrammed between the compression and expansion lines of the diagram, that is, how to increase the effective work. Köhler's criticisms must have caused further doubt concerning isothermal combustion. A third impetus was apparently the Krupp Works, which Diesel visited on May 17 and 18 after returning from an Italian vacation by way of Sulzer Brothers, in Switzerland. After producing his new calculations, Diesel wrote Krupp (not Augsburg or Sulzer) indicating the results of his work would greatly reduce the problem of overly large machine dimensions relative to performance.[127] Krupp's concern about this problem may also have stimulated Diesel to make his corrections.

The results of Diesel's recalculations are contained in a 215-page manuscript entitled "Nachträge zur Theorie des Motors Diesel."[128] Later, in September 1893, after engine testing had begun, Diesel added a rejoinder to Köhler's *VDI-Zeitschrift* article and finally, in November 1893, produced a third manuscript, "Nochmalige Betrachtungen über die bei dem Motor zu wählende Arbeitsweise." None of these manuscripts was ever published, though Diesel quoted excerpts in his *Entstehung* (pages 153–156). His rationalization for not publishing the manuscripts was essentially that "no interest existed in divulging to the outside world in what way the purely theoretical investigations of my brochure were transformed into reality."[129] In fact, he did not wish to endanger his patent or his contracts with Augsburg and Krupp.

In his "Nachträge," Diesel indicated that a number of his critics had held that better overall results could be obtained by modifying the Carnot cycle, even though this would mean a loss of thermal efficiency. "However, no positive conclusions were drawn from these many hints on the definitive manner of building engines— the following chapter is dedicated to this question."[130]

From his calculations, Diesel arrived at three conclusions:

(1) *We ought to compress the air purely adiabatically* and not by a combined isothermal-adiabatic process.
(2) *We must not let the expansion be carried out to atmospheric pressure,* since this would cause the cylinder dimensions to be too large. With incomplete expansion, the cylinder will be considerably smaller but with the same power.
(3) *The enlargement of the indicator diagram for one and the same cylinder is very desirable.*[131]

The first conclusion can be justified by the fact that it would make unnecessary the injection of water into the cylinder during compression. Based on his discussion with Augsburg and Krupp, Diesel went on to say that compressions of 90 to 100 atms had to be accepted as an upper limit imposed by the present state of engine building. He was definitely modifying his theory as he became acquainted with the realities and limitations of the technology of his day. Such constraints were also certainly behind his second conclusion. Diesel justified his third conclusion in the following manner:

> The passive resistances of the same cylinder (with its driving mechanism) can be regarded as almost fixed; therefore, the larger the indicator diagram, the larger the active power, and it is not to be excluded that with a larger diagram a greater active power can be expected, even with less thermal efficiency. It is exactly this point on which hinged most of the criticisms of my engine by practitioners and men of science.[132]

Diesel was finally becoming interested in other efficiencies besides thermal. As he admitted elsewhere in his manuscript, he had earlier been exclusively concerned with thermal efficiency, but had realized that a small sacrifice in that area could be compensated for by a gain in mechanical efficiency. *"Thermal efficiency is not, therefore, decisive; we must become clear about mechanical efficiency."*[133] Kurt Schnauffer has observed that Diesel "had finally freed himself from the bonds of a theory that he had so overestimated, and was becoming an engine practitioner."[134]

Diesel next came to the conclusion that the way to achieve a greater diagrammed area, that is to say, more positive work, was by dropping isothermal combustion and adopting constant pressure combustion. "It can be seen a priori that with the same compression, constant pressure combustion results in a much larger diagram area than by constant temperature where the combustion curve sinks steeply."[135] As Diesel summarized it near the end of his "Nachträge": "(1) Combustion at a constant temperature is *completely* excluded. (2) Combustion at constant pressure is the only choice."[136]

Kurt Schnauffer remarks that it is not true a priori that constant pressure combustion gives more positive work. It is the correct solution, however, because its use affords the possibility of utilizing much more fuel than Diesel had originally predicted, thereby getting rid of all that excess air that needed to be compressed and expanded in the cylinder.[137] He apparently came to this further insight in September 1893, as he was making calculations for the first reconstruction of his test engine.[138] What was needed was about eight times as much fuel as his original theory had called for. The question of the correct air-fuel mixture needed for proper combustion could now be addressed.

Despite these changes, Diesel kept the basic ideas of high compression of air and injection of fuel into the highly pressurized air. He realized now that combustion temperature would rise substantially and by the fall of 1893 had recognized the necessity of cooling his engine. As early as June 1893, and again in the fall of 1893, he calculated that the compression temperature would be between 500° and 600°C, and that compression would be between 30 and 40 atms. These calculations were essentially correct.[139]

Diesel closed his November 1893 manuscript, "Nochmalige Betrachtungen," by recapitulating the revised working process of his new engine:

(1) Compression to thirty, at the most forty atmospheres.
(2) Combustion at constant pressure as high as temperature relationships will allow, where possible 1600° or 1800°C final temperature during combustion. Within certain boundaries, variation is possible without impairing the economic efficiency, especially lower pressures within the above boundaries.
(3) Water cooling, as with every gas engine.
(4) Air intake to fill the cylinder, incomplete expansion.
(5) As fast a speed as possible, where possible double that of the gas engine.[140]

By November 1893, then, Diesel had completed the corrections in his theory necessary to produce a workable engine. As both Kurt Schnauffer and Friedrich Sass point out, if the impetus for these alterations came from others, it was Diesel himself who made the changes in theory necessary to carry them out in practice.[141]

Unfortunately, Diesel did not see fit to acknowledge his mistakes and proclaim his new theory. He certainly knew his conclusions jeopardized his patent because workable as well as unworkable ideas were combined in its first claim. Diesel decided not to answer Köhler's criticisms in print. As he indicated in letters to both Buz and Krupp, "It is certainly not in our interest through an open discussion to arrive at conclusions that we would best keep to ourselves."[142] What is even more interesting is that none of the companies involved with developing his engine seemed much concerned that the combustion process was different from what Diesel had at first claimed, even though his patent was thereby endangered. Diesel had first announced his theoretical changes to Krupp in a very guarded way: "I have subjected several points of my theoretical process to a closer examination. . . . The results are that through a somewhat altered process, the cylinder [dimensions] can be greatly reduced, and I hope in Augsburg to achieve twice the performance originally hoped for."[143] Krupp did not answer this letter. He carefully followed Diesel's further work, but remarked in a letter of October 30, 1893, "I must emphasize that a comparison between the former process and the one now suggested by you is of less importance to me."[144] If Diesel's industrial backers feared an attack on his patent, as they must have, the available correspondence does not indicate this fact.

Diesel did in fact apply for a new patent in November 1893, which was finally granted in July 1895 as Reich Patent Nr. 82168, "Internal Combustion Engine with Variable Duration of Fuel Injection, Taking Place with Variable Maximum Pressure." Diesel may have waited until late 1893 to apply for a new patent because he wanted to see how his industrial partners would react to his changes in theory. Unfortunately, the original patent draft and most of the correspondence with the Patent Office is no longer extant.[145]

If Diesel's main patent, Nr. 67207, had separated the claims concerning high compression, fuel injection, and constant temperature combustion, as was done in the original patent draft, it would have been easy now to add a supplementary patent protecting combustion at constant pressure. Because, however, the final version of Nr. 67207 mixed all these ideas together in one claim, a supplementary patent was no longer possible. Diesel needed to find a way to protect the constant-pressure process without endangering his first patent. He hit upon the idea of tying the two types of combustion together by means of a process that supposedly regulated fuel injection and thereby increased the area of diagrammed work. As he told Krupp, the essence of his new process lay in a quicker injection of fuel to increase the work performed. Shortening the length of fuel admission, however, does not achieve this purpose. What is decisive is increasing the amount of fuel injected. In the actual patent description, Diesel expressed himself differently. In order to achieve this method of regulation, "the size of the fuel admission period must be variable . . . and the maximum pressure at which fuel is admitted must be variable."[146]

The final version of the patent presented two claims, the first of which reads:

Internal combustion engine of the kind described in Patent Nr. 67207, in which variation in performance is achieved by variation in the shape of the combustion curve, made possible by injection of a simple or mixed fuel spray into the compression chamber of the engine using changing maximum pressures or variable length of fuel injection.[147]

According to both Kurt Schnauffer and Friedrich Sass, when Diesel talked of a variable fuel-injection period, he really meant injecting a larger amount of fuel, which would result in a rise in temperature and complete abandonment of isothermal combustion. The wording he used was meant to disguise the fact that his new process was completely different from that described in his first patent and in his *Theorie und Konstruktion*.[148] Was Diesel deliberately obscuring the truth or did he continue to think of his process as still being essentially constant temperature, from which it was necessary to deviate in reality to achieve a working engine?[149]

Some of Diesel's later comments certainly seem to give credence to this latter view. For example, in a letter of March 3, 1900, to Gustav Zeuner, he said:

I picked the modification of the isotherm . . . on purely practical grounds in order to make cheaper engines for equal performances, not however, because it is theoretically better. . . . When on the one hand today the desired high compression is not yet achieved, and on the other hand, the purely practical question of engine costs rules the combustion curve, still it is certain that the achievement of very high compression and the approach to isothermal combustion represent the best thermal process. The time will perhaps yet come when one will make allowance for larger engines in order to accomplish combustion with less rise in temperature, that is, with less loss of heat through the [cylinder] walls.

Diesel goes on to say that his ideas were recognized as theoretically correct by such men as Zeuner and Schroeter, but, instead of earn-ing any recognition for this, "I was persecuted by a number of critics with almost unbelievable hatred. The deviation from the theoretical ideal, which was unavoidable in practice, was denounced as a crime against science, and the new engine was opposed with moral indignation."[150]

Whatever the truth, Diesel's second patent still did not adequately protect his engine. It spoke of engines of the kind described in Patent Nr. 67207, namely those operating with isothermal combustion. Changing pressures during combustion was never achieved during Diesel's lifetime and was only tried by M.A.N. with U-boats during World War I. Furthermore, the second claim of the new patent referred to a process of achieving the pressurized air necessary for fuel injection through compression of the working piston—a process that has never proved to be feasible. As Kurt Schnauffer says, it is regrettable that a better protection was never found for Diesel's invention. This might have spared him many attacks.[151]

During the summer of 1893, Diesel was about to begin tests on his new engine. Not only had he worked out the corrected theory of the engine and received a patent, but he had also managed to win industrial support for the testing of the first engine. His relationship with German industry and the developmental period of the diesel engine will be discussed in the next chapter.

# CHAPTER 4

## From Invention to the Development
## of a Working Engine

•

### Diesel Finds Industrial Backing

Although Diesel had worked essentially alone on his earlier inventive efforts, he must have realized that the construction of his new engine would be too complicated and costly for such an approach. In an effort to obtain industrial backing, therefore, he turned first to his employer, Carl Linde, in a letter of February 11, 1892, which described his theory and suggested that Linde use his influence to interest several factories, including the Augsburg Engine Works, in developing his engine—at their own risk. Diesel also requested financial help from Linde in return for a share in the presumably forthcoming profits from the engine.

Although he found Diesel's theory correct and said many good things about his ideas, Linde refused to back the project because he felt the practical difficulties would be too great. He did, however, offer to try to help interest other firms in the project. Kurt Schnauffer refers to Linde's reply as "peculiar" (*eigenartig*),[1] and one would have to agree with this judgment in view of the mixture of positive and negative comments. If Linde thought the practical difficulties were so serious that he wanted no part in the scheme, why then did he offer to try to interest other firms in it? Was this done as a sop to Diesel's feelings? Although Linde apparently did make some efforts to fulfill his promise, it is impossible to tell how seriously he pursued them. In any event, he and Diesel remained friends, and Diesel certainly was not averse to excerpting Linde's favorable comments and using them in letters to other men and firms he hoped to influence.[2]

Although Linde's response was not what Diesel had hoped for, the example of Linde's own business enterprise surely served as a model for what Diesel sought and would partially achieve in the way of industrial backing. When in 1876 Linde had been perfecting his refrigeration techniques using an ammonia compressor, he had set up a consortium that financed the construction of his machine in return for part interest in his patents. Among the participants were Heinrich Buz, head of the Augsburg Engine Works, and the Krauss locomotive factory, in Munich.

The refrigerator was developed in Augsburg, and, when it was judged marketable in 1877, Linde's consortium entered into contracts with a number of leading machine factories, again including Augsburg, giving them the right to construct and sell refrigerators in certain areas of Germany as well as abroad, in return for the payment of licensing fees. Such agreements included Sulzer Brothers for Switzerland, Italy, and Spain, and Carels Frères for Belgium and Holland. Diesel would later sign contracts with these same companies. Finally, in 1879 Linde created the Society for Linde Refrigerators (Gesellschaft für Lindes Eismachinen), at Wiesbaden, in which Augsburg and Buz again participated. For renumeration and security for his family, Linde gave up the remaining rights to his patents. He was, however, the company's director until 1890.[3]

The close ties of Linde's enterprises with the Augsburg Engine Works and its director Heinrich Buz proved to be an important precedent for Diesel. Linde's arrangement with his Wiesbaden Society contained a number of features similar to Diesel's General Society for Diesel Engines (Allgemeine Gesellschaft für Dieselmotoren), set up in 1898.

Linde continued to make some efforts to help Diesel during the crucial months of March and April 1892. On April 16, after hearing that Augsburg had rejected Diesel's first approach, Linde wrote that he deeply regretted the situation. He promised that, after returning from a trip to the South Tyrol, he would both speak to Buz and eventually would be active in trying to form a consortium.[4] He, however, was still away when Buz changed his mind and decided to support Diesel. It is not known whether Linde subsequently tried to influence Augsburg. On April 20 and June 10, Diesel wrote to Linde asking him to continue his efforts in establishing a consortium inasmuch as Augsburg's agreement was in such a guarded form. Again, the extent of Linde's activity is unknown,[5] but on June 30, 1892, he had to inform Diesel: "I can remember no time in which such a disinclination existed against new industrial undertakings or in which there was such an effort to take one's capital out of industrial enterprises and put it into secure investments."[6]

Linde's comments about the German industrial situation are a clue to the difficulty Diesel faced when he began trying to interest firms in his engine during a recessionary period in German industry. The period 1873–96 has been described by some historians as "the great depression," characterized by alternating periods of recovery and relapse. Although recovery occurred during the period 1888–90, a relapse took place in the four-year period 1891–94, during which stock prices fell and unemployment rose. Not until 1895 did a general recovery get underway.[7] Indeed, Linde's impressions of the period are borne out by the following comments of the social historian Hans Rosenberg:

> Among those investors, both large and small, who in many cases had become anxious and shy of speculation, there was in the period of slackening entrepreneurial desire no lack of elements that preferred to invest their liquid funds in firm, interest bearing familiar government paper or foreign stocks.[8]

Diesel thus found himself operating in a poor business climate.

The same conditions that affected German business in general also affected the Augsburg Engine Works, the firm that would eventually first back Diesel, in particular. It experienced an economic slowdown, beginning in late 1891. Orders dropped off, and 280 men, or 14 percent of the work force, were laid off between June 1891 and February 1892.[9] Between 1890 and 1892 the value of the Engine Works' stock plummeted, and dividends declined in 1892–93.[10] Josef Krumper, head of the steam-engine department, noted in his memoirs that his production fell in 1891–92 by almost half in terms of horsepower output because of the poor situation in the textile industry.[11]

Despite this poor business climate, Diesel next approached Heinrich Buz and his Augsburg firm. Augsburg was certainly a logical choice. Through his connection with Linde, Diesel had been dealing with Buz since 1882, when they had corresponded about manufacturing Diesel's clear ice machine. Furthermore, Augsburg was considered to be a factory of first rank.[12]

Diesel sent his first letter to Augsburg on March 7, 1892, along with his manuscript "Theorie und Construction." He suggested that the theory was essentially sound so that Augsburg need only concern itself with examining the practical side of constructing such an engine. He did, however, suggest that the required pressures in the cylinder would only be 150 atms, not the 250 indicated in his manuscript. Apparently, he had already realized that he needed to modify his theory to make it more palatable to industry.[13]

Almost a month passed before Augsburg answered. In the meantime, Diesel's first patent application had been returned on March 15. Linde's letter of March 20 had declined to back Diesel, and on March 29 Schroeter had written his first noncommittal letter. Then, on April 2, 1892, Buz and the Augsburg factory wrote Diesel, rejecting his proposal. The short communication read in part: ". . . we regret to inform you that we cannot consider the building of the engine in question. We have carefully considered all aspects of the matter and think that the difficulties in execution are so great, we cannot risk it."[14]

So much bad news in the space of about three weeks must have depressed Diesel. However, he did not give up his efforts, which now took two forms. He attempted (as it turned out, unsuccessfully) to interest other firms, especially the Gasmotorenfabrik Deutz, and he sought to change Buz's mind through a series of letters in which he modified his ideas toward more realistic goals—an effort that ultimately succeeded. Diesel apparently convinced the French internal combustion engine expert Gustave Richard, of Paris, to smooth the way for him with Eugen Langen, at Deutz. On April 13 Diesel wrote to Langen, but the latter rebuffed him in a reply of April 19. Langen's letter is interesting, especially because it was he who some thirty years earlier had teamed up with Nicholas Otto to produce the gas otto engine. Now he had a second chance to help pioneer a new form of internal combustion engine. Times had changed, however. Langen wrote to Diesel:

> I do not hesitate to agree with the judgment of Herr Linde and others that your viewpoint is theoretically correct; however, I must add that it essentially contains nothing new, since as physics it is already known and can be found in various known constructions, especially the continual addition of fuel during the combustion period.
>
> Your foreign patents, in so far as they have been granted without preliminary examination, have a limited value, and I suspect that your application in Germany will not lead to a patent.

This was often the case, Langen claimed, in trying to work out theories that are good in themselves. He ended by saying he would be ready to enter into negotiations should Diesel receive a German patent.[15] Diesel would correspond with Langen again in early 1893, but the results were equally negative. The reasons for Langen's attitude will be discussed later in this chapter.

During April 1892 Diesel also corresponded with Mannesmann-werke, in Berlin, but this also came to naught because Mannesmann would do nothing without Langen.[16] While attempting to win over Langen and Mannesmann, Diesel also tried to change Buz's mind. On April 6 he wrote to him attempting to find out exactly what the difficulties were that had influenced the Augsburg Engine Works to reach its negative decision. Diesel must have guessed that the firm was put off essentially by the high pressures mentioned in his first letter and manuscript. Without giving any specifics, he now indicated that his manuscript had only described the most perfect system, yielding the best thermal efficiency and that a reduction of pressures would still result in a large savings in fuel. He also suggested using gas or oil as a fuel, rather than powdered coal. He especially asked for a meeting with Buz in which he could answer any objections of Augsburg's experts.[17]

On the 9th of April, Diesel wrote a second letter, this time mentioning that thermal efficiencies of 68 percent could be reached with a pressure of only 44 atms, one that was within the range of contemporary technology. As Schnauffer says, Diesel was well aware by now of the mistake he had made in originally proposing only such high pressures as 250 and 150 atms. It is difficult to see how he could have expected any factory to seriously consider his project under such circumstances.[18] He also at this time stressed the use of fluid fuel instead of coal. His closing argument turned to economic matters:

> I must once again emphasize that it is not a matter of building a new system of gas or petroleum engines; in that case your hesitation to take up the matter would be completely clear. It is much more a matter of a substitute of something much more simple and complete for the presently constructed steam engine, along with its boiler. Seen from this standpoint, the matter would seem of great—even commercial, or especially commercial—interest for a firm of your world reputation.[19]

Thus, Diesel was not only trying to appeal to Buz's commercial sense, but was also suggesting that his system might replace the steam engine—one of the mainstays of Buz's factory.

Despite these two letters, Diesel still received no answer, apparently because Buz was away on a trip. On April 13, therefore, Diesel sent a letter to Professor Schroeter, trying to answer some of his objections. A copy of this letter was sent on to Augsburg. Here again, Diesel emphasized that his manuscript described only the ideal type and that in practice it would be possible to lower pressures to 44 atms and still reach 67–68 percent thermal efficiency. He then mentioned a number of existing machines that worked with such pressures. He also indicated several possible modifications that might improve his engine.[20]

These letters had their effect, for on April 20 Augsburg reversed itself and agreed to test the engine under restricted conditions— even before a patent had been granted! This all-important letter read in part:

> After your new expositions we wish to declare ourselves ready, circumstances permitting, to undertake the completion of a test engine, which, however, must be of such a design that we can avoid as much as possible all construction difficulties. This should only be the first step to determine whether the system is practically feasible, without at first realizing a significant increase of the efficiency vis a vis other engines. If then the first difficulties, which are always existent, are overcome, one can go forward step by step in small increments.
>
> We leave it to you to submit construction drawings for such an engine and would then invite you to further personal discussions.[21]

Although this letter was written in a very guarded form, it marked a major success for Diesel. One of Germany's most prestigious engine factories had agreed to test his ideas. Throughout the rest of 1892, Diesel and Augsburg corresponded, as drawings and specifications for the new engine took shape. Not until February 1893, however, did the two parties sign a contract to test the engine, after the patent had been granted and Krupp was being won over as a partner in the endeavor.

What exactly transpired in March and April 1892 in Augsburg is impossible to say for certain. The minutes of the board of directors are silent on the decision, and no definitive materials from this period exist that would clarify the situation. Diesel apparently thought that both technical and business considerations played a role. On the original copy of Augsburg's rejection letter of April 2, 1892, Diesel drew a cross with blue pencil and added "*Gutachten Krumper,*" that is, "in the estimation of Krumper." Josef Krumper was the head of the steam-engine division of the Augsburg works, and Diesel was assuming that his expert opinion, among others, had been solicited by Buz and was primarily responsible for the rejection of his offer. In addition, Diesel's letter of April 6 to Linde, in which he reports Augsburg's rejection, contained the following passage:

> I am surprised that the Engine Works refused my offer without having directed a single question to me and without any conversations at all—I must conclude that their refusal was caused less by technical reasons than by Augsburg's disinclination for business reasons to take up anything new.[22]

Later in the letter, Diesel urged Linde to help him build a consortium in case Augsburg's attitude hardened. Diesel's letter of April 9 to Augsburg tried to play to its economic concerns and may well have helped change its mind. Possibly Buz and Augsburg thought they had better test Diesel's theory in case it was correct. Much of their business was in steam engines and they could not afford to take the chance and not test his concept, just in case his comments about dethroning the steam engine proved to be correct.[23] Buz had already been involved with Linde's refrigerator. He had indicated as far back as his first business report of 1865 that engine factories more than any other needed to keep up on the latest improvements in order to withstand growing competition.[24] That this economic argument contained some truth can be surmised from one of Diesel's comments to his wife in a letter of March 6, 1894, after the testing period had gotten under way: "Just think, since the engine has started running, the stock of the Engine Works has risen by 15 percent."[25]

Kurt Schnauffer asserts that Lucian Vogel, Buz's son-in-law, could not have been the moving force in changing Buz's mind, as Eugen Diesel maintains. Vogel's letter to Diesel of June 14, 1892, makes it clear that he had been out of the country on a trip for health reasons and had not yet thoroughly examined Diesel's manuscript. Further, the fact that both men used the formal "Sie" with one another and did not start using the more personal "Du" until 1894 would indicate they were not old school friends, as Diesel later asserted.[26]

Schnauffer is wrong, however, when he attributes the change of heart to Buz alone, based on Diesel's eloquence and on economic considerations. Buz was also influenced by Krumper. Diesel was always suspicious of Krumper, did not regard him as a friend, and so states in his *Die Entstehung des Dieselmotors* (page 89), where he speaks of his continually negative attitude and sarcastic remarks. In response to such statements, Krumper wrote Buz on November 13, 1913, shortly after Diesel's death, asking Buz to set the record straight. The letter reads in part:

> When Herr Diesel came to us in 1892 with the idea of the "Theory and Construction of a Rational Heat Engine," and I had to express my opinion to the Herr Director, I by no means rejected the idea, since the proposal called for better fuel economy. Although the difficulties in execution seemed extraordinary, I interceded for the construction of a one-cylinder engine.
>
> Since we had no experience at all with engines such as gas or oil engines, the first construction of the new internal combustion engine naturally fell to the inventor.
>
> From the beginning I had to excuse myself from collaboration in construction and tests, not out of aversion or stupidity . . . but because of an overload of work at that time.
>
> In this connection I might recall that in the 1890s I led a steam engine and turbine design bureau . . . of twenty-seven to thirty men, carried out a great part of the correspondence, conducted numerous tests in the Engine Works, and in the years in question spent forty to sixty days a year away on business trips.
>
> The comments of Diesel concerning my visits in the laboratory

[Krumper's supposed sarcastic remarks] can only be understood as an argument for the martyrdom he seems to have desired . . . they would have been better left out, since the picture of Diesel's character is not enhanced by them.[27]

Krumper essentially repeated these assertions in his unpublished memoirs of 1916. He stated:

Diesel's long theoretical treatise was not convincing in terms of surmounting the extraordinary difficulties involved. However, since attempts to increase the efficiency of fuels had been in the air for years, Krumper advised the management to agree to Diesel's plans, especially since aside from powdered coal, oil was to be used for combustion . . . K. had to absent himself from close collaboration on the engine and tests because of overburdening in his other work. Since R. Diesel in his 1913 *Die Entstehung des Dieselmotors* speaks very derogatorily of K. and his continual rejection of the engine, I hope that the events are set straight here, events without which Diesel would most probably have not realized and been able to continue his invention at the Engine Works.[28]

In a letter of reply to Krumper, Heinrich Buz bore out these contentions:

One cannot speak at all of a "continual rejection" of Diesel's invention by you, since you not only advised the undertaking of a test engine *from the beginning,* but also in later years came out for the continuing of the tests.

It seems that Diesel was angry that you refused him direct collaboration. But I can confirm that this would not have been possible for you because of overwork in steam engine and turbine construction. It would also seem that Diesel, by emphasizing the lack of support from the steam engine division, wanted to play up his own contributions.[29]

In evaluating this evidence, one must allow for the fact that time had passed since the events described. Krumper may have been exaggerating for effect and trying to take credit for more than he had in fact done. By the time of Diesel's death, his relations with Buz had deteriorated, even leading to lawsuits. Buz, therefore, may not have been particularly anxious to place Diesel in a good light. On the other hand, Diesel was often suspicious of those who worked with him; indeed, he seems to have eventually fallen out with many of those who had most to do with his engine: Buz, Vogel, Lauster, and Krumper. Many of the later memoirs of Diesel's former colleagues are undoubtedly colored by such animosities. Opponents of Diesel and his engine certainly existed, and apparently some men tried to gain through taking out patents on improvements whose authorship was disputed or through dubious financial and stock transactions. Nevertheless, in many cases it is difficult to tell whether real opposition to Diesel existed, whether he imagined it, or whether he created it by unfounded suspicion.

If Buz's testimony is essentially sound, however, Krumper must be credited in large part with advising Buz to back Diesel. At first, both men probably felt the practical difficulties were too serious to undertake the venture, but, after Diesel's further letters mentioning lower pressures in the cylinder, Krumper must have advised Buz to proceed with the tests. Logically, Buz would have listened to the advice of the head of his steam-engine department because, as Krumper points out, the Engine Works had no experts in internal combustion engines. Interestingly enough, this was also largely true at Krupp. Diesel may have been materially helped by the fact that there was no one with substantial experience in engine design who might have urged rejection of his ideas. After all, he himself was no expert in such design.

Later, Diesel would go out of his way to ensure Krumper's goodwill. In a letter of October 13, 1892, he asked for his support during the testing period, and on January 6, 1893, he sent him a copy of his newly published book.[30] If Diesel's later relations with Krumper were not always the best, the latter's service in winning acceptance for Diesel's ideas must be recognized.

In April 1897 the diesel construction bureau at the Engine Works was placed under Krumper's supervision when Buz's son-in-law, Lucian Vogel, left Augsburg, feeling he was not receiving enough credit for his part in advancing the diesel engine. Diesel was not happy about this move, and years later it may have occasioned his negative remarks about Krumper in his *Entstehung* book.

Immanuel Lauster was a young engineer who came to work for the Engine Works in early 1896, was assigned to the diesel engine division, and more than anyone else was responsible for turning the diesel engine into a marketable product. Unfortunately, Diesel and Lauster did not get along, and Lauster's later unpublished work on the history of the diesel engine pictures the inventor in a poor light. Lauster maintained that, when he first came to Augsburg, he gradually became aware that Diesel and Krumper had clashed and that Buz had gone ahead with the original tests in 1893 over Krumper's objections. Lauster does not say, however, where he obtained this information, and his chief proof seems to be that Krumper was not put in charge of the diesel engine tests during the period 1893–97.[31]

Diesel was, of course, extremely fortunate to win the backing of an industrialist like Heinrich Buz. Whatever Buz's original doubts were, he rapidly became the chief supporter for the development of the engine. Without such backing, Diesel and his engine could probably not have surmounted the difficult years that lay ahead. Diesel made this point several times. For example, on the occasion of Buz's retirement from the leadership of M.A.N. in 1913, after his relationship with Diesel had cooled, the latter wrote a letter in which he expressed his thanks: "You were not only the first to recognize the significance of my discovery, you also held fast to it during the creation of the engine, unperturbed by seemingly insurmountable difficulties."[32]

Further testimony to Buz's key role are the following excerpts from Colonel E. D. Meier's report to Adolphus Busch of October

4, 1897. Meier was Busch's chief engineer and had been sent to Augsburg to examine the diesel engine before Busch purchased the rights to Diesel's American patent. Meier says of Buz:

> Commerzienrat Buz, the general manager of the Augsburg Machine Works, is known as the Bismarck of German Machine Industry. Mr. Buz is a man of commanding presence, great firmness and decision. He takes no active part in design or construction, but watches and controls every important detail of his large establishment. As he builds many and excellent steam engines, his firm conviction that the Diesel Motor is destined ultimately to supplant the steam engine entirely, deserves great consideration. It is due to him, above all men, that Diesel was enabled to embody his theories in practice.

And later:

> Mr. Buz, although not an enthusiast by nature, but a strictly practical man of affairs, thoroughly believes that the motor is the greatest advance ever made in dynamic engineering. . . . He has believed in it since Diesel's first essays in 1893, and has stood in the breach with his advice and the power of his name. Mr. Diesel reveres him as a father, and told me that without his unwavering faith and support, he could never have made a practical success of the motor.[33]

By December 1892 Diesel's first patent had been granted, and he had decided to publish his theoretical manuscript with revisions. As he explained in a letter of October 12, 1892, to Augsburg:

> My intention is chiefly to interest the scientific world in my business and at the same time to provoke a judgment from public opinion. It seems to me this course of action can only be advantageous in a business sense, by getting the public used to the new thing, which will save a lot of time later. . . . Finally my new paper contains various variants and combinations which could not be included in the patent publication but whose publication would make impossible

their patenting by a third party. I believe that the publication of the treatise can only strengthen the patent.[34]

Diesel was definitely interested in publicizing his work with the ultimate aim of gaining more industrial support because he still had not signed a contract with Augsburg. Also, Augsburg would only underwrite the tests and Diesel knew he would need to leave Linde's employment. He therefore needed someone to pay him a salary during the testing period.

The responses of academicians and industrialists to Diesel's book have been mentioned in chapter 3. Although a number of leading academicians expressed favorable judgments, industry was more reserved. As was to be expected, objections were raised primarily on practical, rather than on theoretical, grounds. Wilhelm Ochelhäuser, of the Continental Gas Company, argued that the opinions of academicians were of no account because construction of a working engine was a purely practical problem to be solved by practitioners.[35]

Now that a patent was in fact granted, Diesel approached Eugen Langen once again for support. On January 2, 1893, he wrote him that now not only the process but also the engine design were patented. As in other letters of this period, Diesel mentioned four advantages of his engine: (1) it would use only one-sixth to one-tenth of the fuel of the steam engine; (2) the dimensions would be smaller and the engine would not require a boiler or firebox; (3) all forms of fuel—solid, gas, or liquid—could be used; and (4) the engine was not a new type of gas or petrol engine but was meant to replace the steam engine.[36]

Langen's reply was no more positive than the year before:

The matter has found my whole attention, and I and my colleagues in the Gasmotorenfabrik Deutz hold unanimously that what you are striving for is certainly theoretically correct. I hope it won't anger you if I, as one with practical experience, have considerable reservations about your ability to realize these ideas. Experience hardly exists with machines that make 300 revolutions, bear 200 atmospheres of pressure and in a hardly measurable time take in and consume solid fuels. I don't believe I err when I assume that such attempts will involve a huge disappointment. As great as my recognition for your theoretical work is, just as great are my reservations about its practical realization.[37]

Diesel's letter of January 14, pointing out that pressures of 44 atms could be used and still yield major fuel savings, went unanswered. A meeting between the two on the 29th of January in Berlin had no practical outcome—possibly because Diesel was now negotiating with Krupp and felt less need for Deutz's support—nor did a further exchange of letters in May and June.

Thus, Langen missed an opportunity to be involved a second time with a revolutionary new engine. This was his last major decision at the Deutz works, for he died in 1895. Wilhelm Treue says that, by this late stage in his life, Langen was tired, overworked, and weary of the idiosyncrasies of inventors and the patent law (the Otto patent had recently been overturned). Also, times had changed. In 1864 the scientifically uneducated inventor and the pioneer entrepreneur type were complementary figures in German industry, as witnessed by the Otto-Langen relationship. By the end of the nineteenth century, however, the inventive process was becoming more regularized, especially through research labs in industry and in the Technische Hochschulen. Langen had grown used to the new situation and did not wish to repeat his experiences of the 1860s.[38]

Despite the conditions peculiar to his situation, Langen's letter does indicate the chief fear of the industrialists: the seemingly formidable difficulties and risks involved in developing a practical engine. Unlike Buz, Langen possessed much experience with internal combustion engines. Thus, he was much more cautious than Buz and much more aware of the difficulties that Diesel's engine presented. Similar hesitations can be seen in a letter from Crossley Brothers, in Manchester, England, to Bryan Donkin, who was trying to interest English firms in Diesel's engine.

From what we have seen of this engine we should hardly think it worth the trouble of experiment. If the patent has not been based on experiment, we should hardly care to trouble further about it. On the other hand, if Mr. Diesel has made such an engine, and has it working, we should not object to trying it here with a view to a further consideration of the question.[39]

While Diesel attempted to interest other firms in his invention, he was undertaking contract negotiations with the Augsburg Engine Works, a process that lasted from December 1892 to February 1893. Numerous contract drafts were drawn up, but Diesel drove a hard bargain and was not satisfied with them. On February 4 he tried to hurry Augsburg along by indicating that "a great number of interested parties have contacted me and offered me significant sums of money for the German monopoly, with the exception of the South German states reserved for you."[40] In fact, Diesel at this time was only negotiating seriously with one other firm, the Krupp works, in Essen.

Diesel's first letter to Krupp, on January 19, 1893, accompanied by a copy of his book, listed the advantages of his new engine over the steam engine (but did not mention various possible fuels), quoted the favorable opinions of Linde, Schroeter, and Zeuner, and ended by emphasizing that the engine would be made of steel, which "may well be of interest to you."[41] Krupp's answer was positive and led to a meeting between Diesel and Krupp director Asthöwer on January 31 in Essen. That same day, Diesel drew up a contract draft and sent it to Krupp. On February 2 Diesel also forwarded a copy of a draft contract with the Augsburg Engine Works and admonished that the contract signing was only being held up by a few formalities and that, if Krupp signed now, terms might be better than if the firm waited until after Augsburg signed. Diesel was obviously playing off Augsburg and Krupp, hoping to win more advantageous conditions for himself.

While Diesel was trying to pressure Krupp into a contract, the Krupp directors were pondering what decision to make, as can be seen from several of their reports to the head of the firm, Friedrich Albert Krupp. On February 4 they indicated that: "In the view of Messrs. Asthöwer and Albert Schmitz, the theoretical foundations of the project are correct, and it is not unlikely that the practical realization will succeed, and that an engine of epoch-making significance will be created which will surpass the steam engine." After discussing Diesel's terms, which were judged to be quite demanding, the report pointed out that, should the project succeed, the profits would be large and a new market for Krupp steel would be secured. A disadvantage of the proposal was that it represented a completely new branch of business for Krupp. Despite the positive opinions of the firm's technicians and various experts, the probability of success was not yet assured. The enterprise represented a leap in the dark that would require significant resources. A handwritten note at the end of the report indicated that director Asthöwer opposed the project because of the risks involved and its speculative nature. F. A. Krupp's marginal note on the report read: "I am also against all speculation. I would only be interested in the question if it is absolutely not in the realm of speculation and if it is of interest to Magdeburg."[42]

To find out more about Diesel's project, Krupp sent two of his directors to Augsburg in early February, and their report to Krupp of February 18, 1893, indicates it was basically Augsburg that tipped the scales in favor of Diesel. Excerpts from this report follow:

Messrs. Klüpfel and Albert Schmitz have been in Augsburg and have received all information desired on the existing construction plans. The Augsburg Engine Works took up the project before the good judgments of technical authorities were put forth, and they have a good opinion of the engine. . . . The gentlemen there are of the opinion that the theoretical foundations of the project are correct and no grounds exist for supposing the practical execution will run up against insurmountable difficulties. This is at least the case for combustion with gas and petroleum, although they are doubtful that direct combustion of pulverized coal in the engine is possible. . . . They have said they would willingly join with us, if we wish to agree with Diesel concerning the exchange of information during the tests and the fixing of sales conditions. We have thoroughly discussed the situation at today's meeting of the directorate and have come to the conclusion that we should not decline the offer. Certainly, a risk is involved. The basis of the invention is theoretical, and practical tests, from which one could conclude with certainty that the engine works and will achieve the expected large fuel savings, do not exist. We are of the opinion that the risk is not too great for us, and the advantages which await us in case of success and which would secure a significant market for steel are so great that taking on the risks is justified. However, the contract must be so worded that during the testing period, we have the right to withdraw at any time, if we are convinced that the desired goal can not be reached.[43]

Even though the contract between Augsburg and Diesel was signed on February 21, negotiations with Krupp dragged on until April. Diesel, therefore, attempted to hasten the process along. For example, on February 18 he wrote Krupp, including among other things, Schroeter's favorable essay from the *Bayerisches Industrie-und Gewerbeblatt*. In the letter, Diesel mentioned that his contract with Augsburg was all but signed and said that Langen had declared himself ready for the possible production of engines, but that he (Diesel) could proceed no further with such possibilities because he felt himself bound by negotiations with Krupp.[44] He pointed out favorable opinions of experts; indicated that, for all the talk of practical problems, no one had mentioned anything specific; and discussed gas as an alternative fuel to powdered coal.[45] In a letter of February 20, he reiterated that it was the modified engine of 44 atms that would be built. Later, enthusiastic endorsements by Reuleaux and Gutermuth were sent on to Krupp.

By March 18, as can be seen from another report of the directorate to F. A. Krupp, the contract was all but worked out. The report emphasized that Krupp would be sharing the risks with Augsburg, a first-rate engine factory. The final contract between Diesel and Krupp was signed on April 10, 1893; that between Krupp and Augsburg, on April 25.

What did these various contracts stipulate? In its contract with Diesel, the Augsburg Engine Works promised to test a 4-hp engine within six months. It was given the licensing rights for the south German states—Baden, Württemberg, and Bavaria, including the Rheinpfalz—in return for which it was to pay a royalty of 25 percent of the sales price of each engine sold. An important clause, also contained in the Diesel-Krupp contract, was that both parties were required to transmit news of improvements to the other. This was the first example of the reciprocity, or solidarity, clause that would appear in all Diesel's later contracts.

In its contract with Diesel, Krupp received the license rights for the rest of Germany, in return for which the company was to pay a royalty of 37½ percent for each engine sold. In addition, Diesel was to receive a salary of 30,000 marks annually during the testing period. The firm reserved the right to withdraw from the contract at any time during the tests.

The contract between Krupp and Augsburg called for both sides to work "hand in hand" sharing improvements and costs of tests. Profits from sales were to be divided equally to help Krupp recoup the salary expenses and to cover the licensing-fee disparity. These contracts would stay in effect until amended in 1897.[46]

In June 1893 Diesel was able to sign one other contract, with Sulzer Brothers, in Winterthur. Sulzer was not obliged to take part in testing the engine, but did obtain an option on the Belgian, Italian, and Russian patents, in return for a payment of 10,000 marks.[47]

By April 1893 Diesel had, through skilled maneuvering and good fortune, obtained all he had hoped for. In a generally poor business climate, with an untested theory and an engine that might possibly pose major practical difficulties, he had won the backing of two of Germany's top industrial firms. Not only construction costs but also a salary of 30,000 marks were to be paid, which would allow him to leave Linde's employ and work full time on the engine.[48] Both firms had agreed to pay high licensing fees for the sale of each engine. It is not clear why Augsburg and Krupp agreed to such high fees; Kurt Schnauffer suggests this was based on the presumably high savings in fuel to be obtained from the engine.[49]

A number of factors produced Diesel's success, all of which played a role in his winning industrial backing. The fact that attempts to improve fuel efficiency of internal combustion engines had been in the air for a number of years provided a context in which he could operate. Also, his ability to sell himself and his ideas played a considerable role. His confident statements on developing a revolutionary new engine that would supplant the steam engine caught the attention of a number of academics like Schroeter and industrialists like Buz. By emphasizing lower pressures and temperatures, he must have overcome some of industry's natural hesitations to test his theory.

Diesel was also lucky to attract industrialists like Buz and later the Krupp officials. In this connection, it is interesting that representatives of the older, established, and possibly threatened steam-engine technology did not prove to be more of a restraint. Furthermore, the poor business climate of the early 1890s may have been as much a help as a hindrance. If many firms were unwilling to risk a new venture, Augsburg regarded a success with diesel engines as a major way out of an economic slump. Diesel certainly played on this theme in his letters to Augsburg. Economic successes as well as economic reversals can stimulate technology and turn it in new directions. Finally, Diesel was helped by the uncertainties of the thermodynamic theory of his time and the lack of experience with the processes of internal combustion engines. Augsburg's early tests were overseen by members of the refrigeration department, and only one expert in gas and petroleum engines from Krupp was involved in the tests.[50]

A word should be said here about Diesel's relationship to the Krupp firm and its officials. In contrast to Augsburg, where Diesel knew and dealt with Heinrich Buz personally, he met Friedrich Albert Krupp only accidentally in 1898 after the development of the engine.[51] Diesel's contact was with a series of Krupp directors and officials, whom he first needed to convince to support his project and then had to continue to satisfy during the years of testing so that the firm would not avail itself of its right to withdraw from the contract. Chief among these officials, who from time to time visited Augsburg, sometimes unannounced, were directors Asthöwer, Albert Schmitz, and Ludwig Klüpfel; chief engineer of the technical-engine bureau and later member of the board of directors Gisbert Gillhausen; and chief engineer of the Grusonwerk gas-engine division Hermann Ebbs. Of these men, Ebbs was the only expert in gas and oil engines.[52] Unfortunately, throughout the test period, Krupp did not assign anyone to be constantly present in Augsburg, who might have acquired some expertise and who might have helped solve some of the many problems that arose. Gillhausen, for example, bore a number of other responsibilities and could not devote much time to the engine.[53]

Diesel pointed out in a letter to his wife that Gillhausen and Ebbs were the officials to whom most consideration had to be given.[54] Diesel apparently disliked Gillhausen, thinking he was not a friend of the Diesel cause, and made a number of negative comments about him.[55] Most of Diesel's private comments about his relations with the firm were less than favorable. He complained that he had to travel often to Essen to maintain support for his engine and to combat envy and ill will.[56] How much of this attitude was realistic and how much of it was part of his syndrome of distrust of those around him is difficult to ascertain. It would appear, nonetheless, that the Krupp concern's general attitude toward the engine was less firm than that of Augsburg. From time to time, Krupp's resolve needed to be strengthened by Diesel and Augsburg. By late 1896 and early 1897, however, Krupp officials were sending back enthusiastic reports about the engine's development. In his public statements, Diesel was always careful to thank the firm and its officials for their support.[57]

## The Development of a Working Diesel Engine, 1893–1897

The steps by which Diesel, the Augsburg Engine Works, and Krupp produced the first working diesel engine have been dealt with extensively in other works and will only be briefly described here. During the testing period, an exact journal, including many indicator diagrams, was maintained, though much of this material was kept secret until Diesel's speech before the German Society of Naval Architects in 1912. An expanded version of this speech was published in 1913 as *Die Entstehung des Dieselmotors*, which gives a detailed description of the development of the engine and the problems faced by Diesel and his colleagues during this period.[58] All told, during six different testing periods three models of the engine were built.

For almost a year, between the spring of 1892 and the spring of 1893, Diesel was busy preparing drawings for the new engine. In the summer of 1892, Lucian Vogel, Heinrich Buz's son-in-law, agreed to help with the drawings. Vogel, at that time head of the Augsburg Engine Works' refrigeration division, was to be one of Diesel's main supporters during this period. The first test engine was constructed between March and July 1893. It featured an a-frame design and a single cylinder. It was an uncooled, four-stroke engine, which was supposed to produce 20 to 25 hp. The piston diameter was 150 mm and the stroke 400 mm. Fuel was injected directly by means of a hand pump and a spray nozzle. Only pictures and drawings remain of this engine.

The first series of tests ran from July through August 1893. During this period, only one worker was assigned to help Diesel with the tests. The engine was hooked up to a factory transmission belt and never ran under its own power. Much difficulty immediately occurred with leaking valves and gaskets, so that at first no more than 18 atms of pressure could be attained in the cylinder. Numerous hours were spent testing various materials, many of which were soon worn out by the engine's high temperatures and pres-

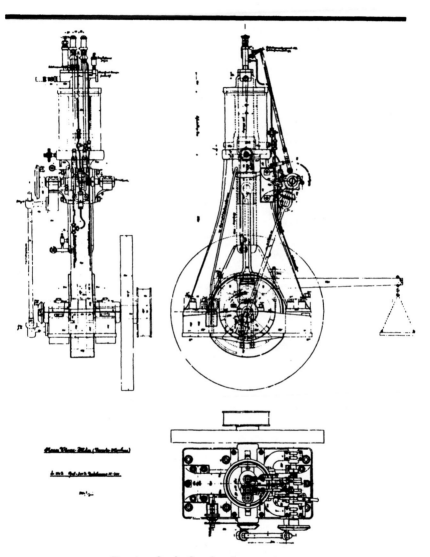

Drawings for the first diesel engine, 1893.
The drawings, based on Diesel's sketches,
were prepared in the Augsburg Engine Works' design bureau.
(M.A.N. Werkarchiv)

First diesel test engine, summer 1893.
The engine was constructed in the spring of 1893
and tested during July and August. Only photographs
and drawings remain of this engine.
(M.A.N. Werkarchiv)

sures. After adjustments were made, a pressure of 33 atms was achieved. Diesel's letters to his wife during this period mirror his impatience with the slow course of events, as he waited to test his basic principles.[59]

On August 10 gasoline (*benzin*) was injected for the first time into the cylinder and immediately ignited, shattering the indicator valve and sending pieces flying past the bystanders.[60] As Diesel was to say later, "After we recovered from the shock, our joy was great, for combustion had proven itself an automatic part of the cycle."[61] He explained to his wife that success was no longer in doubt, "now that the fundamental principles have proven correct."[62] Further tests with direct fuel injection resulted in alternating violent explosions and misfirings, which no one could explain. However, on August 18, Diesel wrote his wife: "The engine made its first independent revolution today, only one, but the principle is proven."[63] Despite these accomplishments, his hopes for a quick success were thwarted. As he said:

> In spite of [the correctness of the cycle] the tests were a failure, since the first machine never ran independently by itself. I returned to my home, at that time, Berlin, very depressed, and made drawings for a complete rebuilding of the engine, which lasted for five months.[64]

These first tests had demonstrated to Diesel some of the problems that needed solution. One major difficulty involved the hardware, such as valves, pumps, and cylinders—what Lynwood Bryant calls the "plumbing."[65] The demands of Diesel's engine were all but beyond the technology of the time, but at least he had a tradition to call on, namely past experience with the steam and earlier internal combustion engines.

The other major problem concerned fuels and combustion. In this area, Diesel was often operating in the dark, and had to move forward empirically because at that time much uncertainty existed about these subjects. He originally thought almost any kind of fuel could be burned in his engine. The very first fuel used, a crude oil, proved to be totally inadequate, and tests proceeded with gasoline and kerosene. The search for the right fuel would last into the twentieth century. Diesel was later to say that the first tests highlighted the most difficult problem of the development period: "the groping search for the correct shape, size, and position of the combustion chamber in the cylinder. This search was connected with the fundamental processes of vaporization [of the fuel] and combustion."[66] In his *Entstehung*, Diesel repeatedly emphasized these key problems of fuel vaporization and combustion—problems that could only be solved by trial and error.[67] Mistaken notions in this area would cost much loss of time.

The rebuilding of the engine lasted until January 1894. The new engine featured an enlarged fuel pump, plus a sheet-metal water jacket for cooling purposes, an indication that Diesel had finally abandoned the idea of isothermal combustion. In addition, the piston and combustion chamber were rebuilt, and the air intake and exhaust valves were separated.

During the second testing period, January to April 1894, Diesel was assigned two refrigeration mechanics, Hans Linder and Franz Schmucker, who alternated in helping him with his work. They would remain with him until the end of the testing periods and would become the first experts in diesel engine building.

Much time was again expended in solving the problems of correct packing for the piston and air-fuel lines. Diesel was especially concerned with the problem of the proper mixture of air and fuel during combustion. A new controlled-nozzle needle was used for fuel injection. A brass coil containing finely bored holes was set in the nozzle's upper part to act as a mechanical vaporizer. This was the beginning of the search for the correct type of vaporizer, a search that would last until 1899. Because direct fuel injection proved to be unsatisfactory, the idea of air-blast injection by means of a Linde air compressor was tried. Diesel probably obtained this idea from a study of the Brayton hot-bulb engine. The engine thereby achieved its first idle running, but Diesel at this point viewed the air compressor as an unnecessary complication. During this period, kerosene was used as a fuel for the first time. As the tests proceeded, Diesel commented to his wife that Augsburg's stock had risen by 15 percent: "That means an increase of the stockholders' wealth anywhere from 300,000 to one half million marks, and I have no part in it, only about 500 marks from a few shares that I possess."[68]

Unfortunately, Diesel still did not know how to obtain the proper mixture of fuel and air during combustion. He decided that the fuel would need to be gasified or vaporized in some sort of carburetor before it was introduced into the cylinder. This, along with experiments using both lighting gas as a fuel and various ignition devices, took up the third and fourth testing periods, June to September and October to November 1894. During these tests, Diesel was willing to abandon the idea of compression ignition—he was willing to do anything to achieve combustion.[69] Although these tests ended in failure, he finally realized that air-blast fuel injection would produce a better mixing of fuel and air and thereby ensure more complete combustion. He also recognized that all the air in the combustion chamber had to be used in the combustion process.

As Diesel said to his wife in a letter of October 3, 1894:

Earlier in the year I got close to my goal much more quickly than expected. Unfortunately, I then followed another direction, hoping for better results. Today I see that the earlier road was the correct one and must be followed. Therefore, I must reestablish the results of February (you were present at the time) and then complete them.[70]

In another letter, Diesel stated, "Buz is the only one that supports the cause through thick and thin and has not a minute's doubt or impatience. He himself has suggested building a new engine."[71]

Construction of the second engine began in November 1894 and lasted until March 1895. Diesel was finally assigned an engineering assistant, Fritz Reichenbach, a brother-in-law of Heinrich Buz.[72]

Also, a special diesel engine design bureau was now set up in the factory, replacing a table in the testing area that had served this purpose earlier.

The piston was rebuilt with an enlarged diameter of 200 mm, and for the first time the relationship between piston stroke and diameter was critically studied. In the first engine, the combustion chamber had been a hollow area set in the piston itself. It was now displaced to the cylinder head. Much attention was paid to minimizing cavities in the combustion chamber, in which air could be lost during the combustion process. Diesel remarked that this concern for the shape of the combustion chamber was the main reason for the engine's reconstruction and further improvement. A fuel distributor, called by Diesel the "star burner," extended from the fuel nozzle into the combustion chamber itself. Although an improvement, it was still not an adequate solution to the problem of vaporization and much time during this period was taken up testing a variety of distributors.

On April 29 actual testing began, resulting in regular operation, good indicator diagrams, and no exhaust noise or smoke. A journal entry on May 1 indicated that the proper indicator diagram had finally been achieved. By May 30 kerosene had been settled on as a fuel. On July 26 the first dynamometer tests were run. The thermal efficiency was 30.8 percent, the mechanical efficiency 54 percent, and the net, or brake, thermal efficiency 16.6 percent.[73] Further work with the piston rings and improved lubrication resulted in a mechanical efficiency of 67.3 percent and a fuel consumption of 327 grams per effective horsepower hour, half as much as other lower compression engines.[74] By September a Linde air compressor had been built into the engine, though at first test results were not as good as those of July. By the end of the year, the engine had built-in air and fuel pumps and was an independently running machine.

Diesel's letters to his wife during the summer of 1895 sound increasingly confident, as though the main work had been completed. As early as June 1895 he wrote to her that the oil-fueled engine was ready for production.[75] On July 11 he mentioned that "a design bureau has been set up, soon the organization for the exploitation of the engine will begin."[76]

Possibly the first opinion from one of Diesel's partners that the engine could now be marketed was the report of Krupp official Gillhausen dated November 5, 1895.[77] This report was based on a visit to Augsburg on October 25 by Schmitz and Gillhausen. Apparently at that time the question of Krupp's right to withdraw from the contract was raised, though this can only be inferred from subsequent letters of Diesel and Vogel. Kurt Schnauffer and, based on his work, Sass maintain that Krupp announced its intention of withdrawing from the contract and had to be won over again by Diesel and Buz.[78] Schnauffer suggests that Krupp was both disappointed that tests with gas had not progressed further and was also fearful that large-scale demand for oil would force tariffs on that commodity even higher.[79] Furthermore, the firm was beginning its own experiments with otto gas engines. Possibly also the less-than-satisfactory tests with the Linde air compressor in early October played a role.

Exactly what the Krupp officials said or how serious Krupp was about withdrawing from its contract is impossible to tell. Certainly, other documents of the period do not make it seem as if a serious threat existed. For example, only two days later, on October 27, Diesel wrote his wife that the Krupp officials were quite enthusiastic and spoke of further tests and the commercial exploitation of the engine. He then added, "Buz thinks it inexpedient to bring up the matter of Krupp's right of withdrawal, since in his opinion there can be no more talk of such a withdrawal—which is also my feeling."[80] By early November, Diesel was asking Krupp to pay his moving expenses from Berlin to Munich, which the company agreed to do.[81] Whatever transpired at the October 25 meeting, Krupp made no use of its right to withdraw from the contract for at least another year, by which time the engine seemed to have been successfully developed.

On January 23, 1896, both Diesel and Augsburg wrote to Krupp suggesting that manufacturing of the engine could begin. Augsburg stated that:

> We believe the time has come energetically to begin exploitation of the engine, now that it has been confirmed that it is significantly superior to every other petroleum engine. . . . Our workshops will not be sufficient in the coming years for the building of engines in large quantities. . . . We suggest that licenses be offered to able firms that already build similar engines, primarily, the Gasmotoren-fabrik Deutz . . . such firms should be invited to examine the engine without any obligation. . . . When the advantages of the engine are recognized, it will not be difficult to negotiate agreements.[82]

Diesel's letter said in part:

> In view of the great importance that is given to this type of engine everywhere and the general attention it has received, I think the time has come to begin commercial exploitation, without changing our testing program in the least. I believe our engine will rule the market and that going public with it is the best preparation for our real goal: the production of large, industrial machines using coal in gas or powdered form.[83]

Diesel went on to suggest that the Krupp plant at Magdeburg might immediately begin producing diesel engines. He repeated Buz's suggestion that Deutz should be won over as a licensee. Clearly, the latter part of Diesel's letter was trying to appeal to Krupp's specific interests. Diesel and Buz may have been pushing the idea of marketability at this point so that Krupp would become convinced and would drop any ideas it might have had about withdrawing from its contract.

Despite both Augsburg's and Diesel's pleas that Krupp approach Deutz, apparently nothing was done immediately. In a letter of

September 16, 1896, Augsburg complained it had heard nothing about the Deutz matter and hoped Krupp would soon contact the firm. This letter evidently was effective. On October 26 Krupp wrote that Deutz had been approached and was quite interested in the diesel engine.[84]

A report of Krupp director Ebbs, made on January 30, 1896, stated that, though the engine deviated greatly from the original plans, its efficiencies were superior to any other internal combustion engine. Only a few details needed changing to begin regular production.[85] The results of these letters were conferences between Krupp and Augsburg officials and Diesel on February 20 and 22, 1896. The decision was made first to build a larger third test engine before taking up commercial exploitation. The second engine continued to be used for testing until it was retired in September 1896. It is today exhibited at the M.A.N. Werkmuseum, in Augsburg.

The third test engine was designed according to Diesel's plans during the spring of 1896, primarily by Immanuel Lauster. Construction was finally completed by October 1896. The piston had a diameter of 250 mm and a stroke of 400 mm. The combustion chamber was finally a unified, even space between piston and cylinder head and lacked any kind of indentations. Diesel, in his *Entstehung*, again called attention to the development of the combustion chamber as the key to his engine's progress.[86] The engine used a single-stage air compressor because he felt this was less complicated than a multi-stage compressor. After a thorough testing of all parts, the engine went into operation in December. The next month, a so-called sieve vaporizer was installed in the fuel nozzle. It consisted of two horizontal disks in which small holes were bored. In between the disks, a fine wire gauze was coiled. Tests with the new vaporizer in January 1897 resulted in a thermal efficiency of 31.9 percent, a mechanical efficiency of 75.9 percent, and a brake thermal efficiency of 24.2 percent. Fuel consumption was 258 grams per effective horsepower hour. Although such results

1897 diesel engine, now displayed
in the Deutsches Museum and billed
as the "first" diesel engine. This was the engine
that was tested and examined by experts
and proclaimed to be a fully marketable product.
(M.A.N. Werkarchiv)

were impressive, in fact the single-stage air compressor and the sieve vaporizer would turn out to be two major trouble spots in the early engines.

By February 1897 the tests had been completed, and interested parties began arriving in Augsburg to examine the engine. It was put through some severe tests, but, as Diesel said, "nothing could influence the quiet, steady working of the engine. It was finally realized that the engine was not only completely developed in its constructive details, but that even using illuminating gas, it represents a 50 percent savings in fuel and cylinder dimensions in comparison to the explosion motor."[87] This third test engine is today in the Deutsches Museum, billed as the "first" diesel engine.

Upon the occasion of the final tests, Diesel wrote to Buz to express his thanks:

> Although I am far from the view that my engine now matches up to the goals I set myself, I should like to say that the results are well above those of earlier engines and that the principles that I represent signify a new era in engine building.
>
> I am completely aware that I could have only come so far through your kind, appropriate and generous support. I also know that you must have had very great confidence in me and my cause in order to promote it as you have and to have never lost patience even in dark moments where one could notice no real progress. You always aimed at the goal with superior wisdom and experience.[88]

Diesel was now at the high point of his career. The three and one-half year difficult period of construction had finally come to a successful conclusion. The engine was about to be officially tested by Moritz Schroeter and announced to the world. Almost all parties concerned were convinced it was marketable.

The fact is that it was not. The over-hasty attempts at marketing would set the diesel engine back for some years and would permanently damage Diesel's fortunes. The process of moving from development to innovation was to be a difficult one.

# CHAPTER 5

## From Development to Marketability: The Early Years of the Diesel Engine

•

"I can say that from today on the difficult time is over and that the rest will develop automatically based on its own worth."[1] Thus did Diesel confidently write to his wife, Martha, at the beginning of 1897. Indeed, prospects must have appeared bright. The engine was now running, apparently satisfactorily, and the delayed phase of innovation, or marketing, was about to begin. Except for one threatened and one actual lawsuit against his patents, the year 1897 was to be an unbroken string of successes for Diesel. Yet, appearances were deceiving, for the years 1898–1900 were to represent a nadir for the engine's fortunes. Only slowly after 1900 did they recover and the engine become marketable. Diesel's own fortunes, however, were never to be restored. The early failure of the diesel engine to achieve marketability is a complex study of the interplay of personal, technical, and economic factors and shows that the road from invention and development to innovation is not always a smooth one.

### Engine Prospects in 1897 and Early Licensing Agreements

In February 1897 interested parties, such as Jakob Sulzer-Imhoof, of Sulzer Brothers, and Frédéric Dyckhoff, the owner of an engine factory in Bar-le-Duc, France, began visiting Augsburg to examine and test the diesel engine.[2] On February 17 Professor Moritz Schroeter performed the official tests. On April 27 and 28 in Augsburg and Munich, respectively, Diesel and Schroeter lectured to audiences of engineers and other experts. The favorable response can be judged by the comments of Professor Alfred Musil, of the Bremen Technische Hochschule:

> Those two days will without doubt become red letter days in the history of our modern machine building, for the success which Diesel has so far achieved with his motor is already so brilliant, the fuel consumption during the perfect working of the machine is so small in comparison to all other heat motors that in spite of all former opposing prophesies, the Diesel motor is not only in full justice entitled to the name of "Rational Heat Motor," but may be fitly called "the Source of Power for the Future." The inventor is entitled to the unreserved acknowledgement and gratitude of his contemporaries.[3]

Finally, on June 16, Diesel and Schroeter appeared before the annual meeting of the VDI in Kassel to present talks on the engine. This was the first official announcement to the world of the results of the long testing period from 1893 to 1897 and was little short of a sensation.

In his talk, Diesel gave some useful background information on how he moved from his ammonia engine to the concept of the diesel engine. Also summarizing the development period from 1893 to 1897, he stated that the thermal efficiency of his engine was between 30 and 40 percent and the net (or brake thermal) efficiency was 26.6 percent, compared to the steam engine's net efficiency of 12 to 13 percent. Although mechanical efficiency was still below that of the steam engine, improvement could be expected soon. He still argued that the compound engine was the preferred way to achieve greater efficiencies and ended by claiming the way was now open for the disappearance of the steam engine.

In speaking of his original theory and its modifications, Diesel

Rudolf Diesel, Heinrich von Buz, and Moritz Schroeter
at the VDI meeting in Kassel, June 1897.
Diesel and Schroeter announced the engine
to the world for the first time. Diesel appeared
to be at the height of his career.
(M.A.N. Werkarchiv)

made it appear as though the main change was dropping isothermal compression so that more realizable pressures could be achieved. He did not mention the unpublished documents that contained the modifications of his theory. Indeed, he maintained that his engine still ran with "approximate isothermal combustion."[4] He probably veiled the truth to protect his first patent, which had less than a year to run before it could no longer be legally challenged (see figure 3).

For his part, Schroeter presented the technical results of his February tests. Unfortunately, he also contributed to two misconceptions about the engine. He made it appear that it was virtually an embodiment of Diesel's original theory. Although he admitted opinions differed about the feasibility of realizing Diesel's ideas in practice, he also said, ". . . from the beginning only one opinion reigned among the experts: that these theoretical principles are completely correct and indisputable."[5] Yet, he certainly knew that

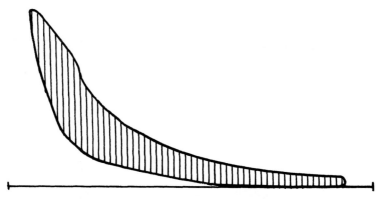

FIGURE 3. PRESSURE-VOLUME DIAGRAM OF THE 1897 DIESEL ENGINE.
The maximum pressure achieved was 34 atmospheres (atms) or 500 psi. Modern diesel engines are capable of higher pressures because of supercharging, but essentially they produce the same diagram. The diesel process is not constant temperature, as Diesel wished, but is a combination of constant pressure and constant volume processes. Drawing by Ellen Thomson, adapted from Diesel's article "Diesels rationeller Wärmemotor," Sonderabdruck aus der *Zeitschrift des Vereins deutscher Ingenieure*, 1897 (Berlin 1897), p. 9.

experts such as Otto Köhler challenged the workability of Diesel's theory, and he must have been aware of the changes made in that theory to produce a working engine. In speaking of the engine's thermal efficiency, he went on to say that it was "the triumph of a theory that could not be more splendidly imagined when one considers that the present realization of the basic ideas is not the end, but that the engine stands at the beginning of its development."[6]

Schroeter may have been trying to say that—because the engine was doing so well in its beginning stages, had gone so far in achieving greater thermal efficiency, and could expect additional improvements—he was justified in calling it essentially the triumph of a theory, even if the theory was not fully realized. Yet, in his conclusion, he once again obscured the issue and made it appear as though theory and practice were one:

> . . . we are dealing here with an engine that is completely marketable and completely worked out in all details. That the thought of the inventor received such a complete embodiment in the short time of three years is due to the same cause from which German machine industry in general has been successful: the combination of fundamental scientific education with untiring energy and constructive ability, which is not deterred by any difficulty.[7]

This quote also contains Schroeter's second error, one shared by many of his contemporaries: that the engine was now marketable. In fact, it was not, and this would lead to considerable difficulty during the ensuing years. Schroeter had been one of Diesel's earliest supporters and one of the leading figures in the movement to replace the steam engine with a more thermally efficient engine. By 1897 his judgment may have become clouded by his ego-involvement in the diesel project. Further, as Colonel Meier, Adolphus Busch's chief engineer, said of Schroeter in October 1897: "Of course he is mainly interested in the scientific side of the subject, and leaves the practical exploitation of the motor to practical and business men."[8]

Interestingly enough, at the very time that Diesel's success finally seemed assured, Krupp was expressing reservations in private to Augsburg about the patents and the marketability of the engine. The occasion for these remarks was threatened lawsuits by Otto Köhler and Deutz, about which more will be said later in this chapter. In a letter of March 2, 1897, Krupp wrote to Augsburg that the firm was worried Diesel's patent might be invalidated, if not by the Patent Office, then by the Federal Supreme Court. Krupp further stated that the testimony of various experts as to Diesel's originality was probably of little worth. In a letter of March 20, Krupp suggested to Augsburg that licensing negotiations wait until the five-year term in which the patent could be challenged had run its course and until the engine was further developed so that more types of engines and more results from actual experience would be available. Krupp suggested carrying out negotiations with Deutz in a dilatory fashion.[9] Apparently, both Diesel and Buz were able once again to allay Krupp's doubts about the patents, so that the company went ahead and signed a new contract aimed at exploitation of the engine.[10]

Krupp, together with Augsburg and Diesel, signed the supplementary contract on March 11, 1897. The contract stated: "Now that . . . a marketable engine of the diesel system is being constructed and tested, the manufacture of diesel engines should begin as quickly as possible." The contract further stipulated that Diesel would reduce his royalties from 25 percent to 5 percent for Augsburg and from 37½ percent to 5 percent for Krupp. Krupp's obligation to pay Diesel 30,000 marks yearly was replaced by a common obligation to pay him at least 30,000 marks annually for the next five years as an advance payment on the 5-percent royalties. The payments were to rise to 50,000 marks a year if either company signed licensing agreements with third parties. All other payments received from third parties for licensing rights were to be shared equally by all three contracting parties. Diesel promised to counsel Augsburg and Krupp concerning further development of the engine.[11]

By 1898 it would become apparent that Krupp's initial hesitations were correct and that a number of improvements needed to be made before the engine was in fact a reliable, marketable product. This miscalculation would send the diesel engine into a period of low ebb, from 1898 to 1900, when licenses would be given up, manufacturing would cease, Diesel would suffer a breakdown and effectively lose control of his product, and the whole diesel enterprise would come perilously close to demise. Later in this chapter, some of the reasons for these misfortunes will be discussed. Here the question should be asked: How could such a miscalculation have been made? Part of the reason may have been the overly optimistic opinions of men like Schroeter. In addition, Diesel and his business partners must have naturally desired to recoup the money spent on the long period of testing and to start realizing some of the supposed profit, especially inasmuch as the main patent had only ten years left to run. If two such prestigious firms as Augsburg and Krupp stood behind the engine, why should not other prospective licensees also be convinced, especially because the engine ran so smoothly on the test stand during their inspection trips to Augsburg?[12]

Because the engine appeared to be marketable, it is not surprising that Diesel and his partners now began to negotiate licensing agreements with a variety of firms. Augsburg had already made it clear it would not have the capacity for large-scale production for several years and had been urging an agreement with Deutz since 1896. For their part, most interested companies and individuals were *at first* quite enthusiastic after inspecting the new engine. The high point in licensing agreements was reached between the spring of 1897 and September 1898, when the General Society for Diesel Engines (Allgemeine Gesellschaft für Dieselmotoren), an organization that took over control of Diesel's licenses and patents, was founded.

Diesel's agreements with the Augsburg Engine Works and Krupp allowed these companies to grant licenses to other German firms. Diesel himself maintained the right to conclude agreements in foreign countries. By 1898 some 87 different patents had been issued in seventeen countries; by 1904 some 141 in thirty-seven countries, with a variety of expiration dates.[13]

Augsburg and Krupp concluded licensing agreements with eight German companies up to the founding of the General Society. The two key firms were Deutz and the Nuremberg Engine Works. The latter firm would merge with the Augsburg Engine Works in December 1898 to form the Augsburg-Nuremberg Engine Works (M.A.N.).[14] Both firms signed contracts in July 1897.

Deutz had been interested in the Diesel patent since Krupp approached it in late 1896 and had examined the engine in February 1897. Apparently to test the Diesel consortium with a view of obtaining better terms, Deutz informed Augsburg in February that it was preparing a legal challenge to Diesel's main patent. Although Krupp was shaken by this announcement, both companies outwardly held fast to the contention that the patent was sound. As Augsburg expressed it in a short telegram to Deutz: "Diesel's patent considered uncontestable. We cannot negotiate on the basis of contestability."[15] Rather than taking on Augsburg and Krupp, Deutz resumed negotiations again, which finally led to a license in July 1897.

The conditions were harsh. Deutz was required to pay 50,000 marks for the license and royalties of 20 percent for engines up to 16 hp and 30 percent for engines over 16 hp. After 1900 the company had to guarantee 20,000 marks annually in premiums. It also agreed not to challenge the legality of Diesel's patents and to accept a reciprocity clause for the exchange of information on improvements. Furthermore, an export license for Russia promised by Diesel to Deutz was never granted. After manufacturing only two engines, the firm called a halt, withdrew from its contract in 1902, and only after 1907 began again to manufacture diesel engines. Also by 1898 Nuremberg had given up building engines. The reasons in both cases were the technical difficulties that were involved during the early innovative phase.[16]

Meanwhile, Diesel had actively been negotiating with representatives of foreign companies. In France, Diesel's personal friend, the manufacturer Frédéric Dyckhoff, had been granted a license in 1894. In April 1897 Dyckhoff founded the Société Française des Moteurs Diesel in Bar-le-Duc. Diesel received stock worth 600,000 francs for his patent rights.[17]

In March 1897 Diesel sold his English patent to Mirrlees, Watson, and Yaryan, in Glasgow, for 100,000 marks and premium payments of not more than 25 percent. Diesel conducted negotiations in Glasgow. Mirrlees was concerned about the patent's validity, but signed a contract after Lord Kelvin gave a good opinion of it.[18] As Diesel wrote to his wife, "They view me here as a very great animal, a sort of Watt, and say so openly to me."[19] Mirrlees would build only one engine before selling the patent back to the General Society.

Diesel also signed agreements with the Swedish banker Markus Wallenberg, who founded the Aktiebolaget Diesels Motorer, Stockholm, in January 1898, from which Diesel received 50,000 kroner and an equal amount in stock. Burmeister and Wain, in Copenhagen, also signed a contract in January 1898, in which they agreed to pay Diesel 20,000 marks and 10 percent royalties.

Several licenses were granted for Russia, the most significant of which went to the Swedish industrialist Emanuel Nobel, nephew of Alfred Nobel, who owned the L. Nobel Engine Works, in St. Petersburg, and who had also developed oil fields in Baku. The agreement was signed in April 1898 and provided Diesel with 600,000 marks cash and 200,000 marks worth of stock in the newly founded Nobel Society, in Nuremberg, which was to handle all Russian licenses. Diesel commented to his wife:

Note well this date [February 16, 1898]. It is the date of the signature of my association with the petroleum king Nobel and probably the beginning of world shaking events. The cold Swede is now more than myself fire and flame for my engine. Let me dispense with the details. I am almost completely exhausted and can do no more after such a settlement that requires daylong straining of all faculties.

After discussing the financial settlement and speculating on the money he would make if his stocks increased in value, he suddenly drew away from such thoughts and commented, "There, I sound like those awful money Jews. Let's forget it!"[20]

Diesel's most lucrative licensing agreement was, however, with Adolphus Busch, a German-American and founder of the Busch Brewery, in St. Louis. Busch was spending several months visiting German health resorts in the summer of 1897. He had been made aware of the new invention by his chief engineer, Colonel E. D. Meier, and by the hops merchant Berthold Bing.[21]

Bing and Meier were both early converts to the diesel cause and were to play important roles in the early history of the diesel engine. Colonel Meier (1841–1914) was of German ancestry, though born in St. Louis. He had studied four years at the polytechnic school in Hanover and was fluent in German. After fighting for the North in the American Civil War, he had worked for a number of railroads, owned several companies that made engines and boilers, and then gone to work for Busch. He was to become a lifelong friend of Diesel and was one of the pioneers of the diesel industry in America.[22]

Berthold Bing (1847–1915) owned an established merchant and wholesale firm in Nuremberg that supplied hops for Busch's brewery. His business sense was a good complement to Diesel's often unrealistic expectations. Bing helped in negotiations with Wallenberg, in Sweden, and in setting up the British Diesel Engine Company in 1901. Diesel had also entrusted Bing with the creation of the General Society in 1898.[23]

Busch's first meeting with Diesel took place on September 6, 1897, in Baden-Baden, out of which came a preliminary agreement. By September 9 Busch was in Augsburg to see the engine for himself, and as a result he instructed Colonel Meier to travel there to test the engine and to interview relevant persons for their opinions. Meier arrived at Augsburg in late September, and, because of a highly positive report he submitted on October 4, Diesel and Busch signed a final contract on October 9. Busch received the right to the Diesel patents in the United States and Canada, in return for which he paid 1,000,000 marks and agreed to a royalty of 6 percent per engine. In January 1898 Busch founded the Diesel Motor Company of America in New York City. After such a promising beginning, the early history of the diesel engine in America was to be extremely disappointing.[24] Diesel, however, was now a millionaire, and his fortunes, both literally and figuratively, seemed to be at their height.

Colonel Meier's report of October 4 was extremely interesting because it gave a good picture of the personalities involved with the diesel engine in the fall of 1897, their opinions, and how the future of the engine must have appeared at that time. In his introductory remarks, Meier mentioned the low efficiencies of the steam engine and Sadi Carnot's theory of the perfect heat engine. "This," he said, "has since then been the ideal of mechanical engineers."[25] Meier credited Zeuner and Schroeter with being the leading exponents of thermodynamics in Germany and leaders of the movement toward more efficient engines. After discussing various men he interviewed and their opinions, such as Anton von Rieppel, head of the Nuremberg Engine Works, who thought the engine adaptable to all forms of vehicles, Meier concluded: "The general impression carried away by me after the interview with all these men was that they are sincerely and earnestly convinced that we have before us a great and important invention, whose possibilities it would be hard to overestimate."[26]

As to what contemporaries thought the essence of Diesel's invention to be, Meier said:

Lord Kelvin considers "Diesel's invention of introducing the fuel into air previously heated so far above igniting point merely by compression, and so causing the fuel to ignite, thoroughly original. . . .

In nearly all previous gas and oil engines the air and fuel are compressed after being mixed together, with the great disadvantage that the degree of compression cannot go beyond comparatively narrow limits without causing explosion before the compression has ceased." In this lies the gist of Diesel's invention.[27]

Meier then discussed the various licensing contracts that had already been concluded or were in the process of negotiation and indicated strong activity in this area because Diesel was receiving new inquiries almost daily. In several places, Meier called attention to Diesel's insistence in his contracts on sharing improvements among licensees through the reciprocity, or solidarity, clause and believed this was a major reason why the diesel engine would make rapid progress.[28] His description of Diesel was also flattering. He contended that all of his success had not gone to his head and that he was still a modest and simple man whose amiability promised to foster harmony among various diesel interests.[29]

Meier believed the diesel engine would soon surpass the steam engine in small horsepower uses and in the not-too-distant future would be adaptable to marine transportation: "The broad minded and liberal policy inaugurated by Mr. Diesel in making all licensees agree to interchange experience and inventions, insures a rapid development in every field now commanded by the steam engine."[30]

Meier's report concluded with the opinion that Diesel's patents were strong. "I believe that the purchase of the Diesel patents for America is as promising an investment as the purchasing of any patent claim could be." Applying Diesel's method for granting licenses that had worked so well in Europe "would soon create from among the largest and best engine works of the United States, an interest so strong that no one would dare to attempt infringement or evasion of our patent."[31]

Meier was certainly one of the strongest supporters of Diesel, but his comments are probably fairly typical of this period of early enthusiasm for the diesel engine and help to explain why Diesel was able to negotiate so many patent and licensing agreements. Two other companies that Diesel was involved in, the Diesel Engine Company of Augsburg (1897) and the General Society for Diesel Engines (1898), will be examined later as case studies in what went wrong with his affairs during the early years of commercial exploitation.

By the spring of 1898, Diesel had achieved considerable success in exploiting his numerous patents, though the storm clouds were gathering. He was already beginning to experience difficulties with his colleagues. For example, Lucian Vogel, Buz's son-in-law and one of Diesel's early supporters, had left Augsburg in the spring of 1897, feeling he was not receiving enough credit for his role in developing the engine nor being entrusted with important enough tasks. Vogel apparently was not easy to get along with, and Buz was rather brusque, even with his son-in-law. In February 1898 Buz's brother-in-law, Fritz Reichenbach, Diesel's first engineering assistant during the testing period, and Rudolf Pawlikowski, the head of Diesel's private design bureau, in Munich, were both dismissed over disputes concerning priority in patenting certain engine improvements.[32]

Nonetheless, in the spring and summer of 1898 Diesel saw two further triumphs. In May 1898 a diesel engine, constructed by the Nuremberg Engine Works and bought by the Diesel Motor Company of America, was exhibited at the Electrical Exposition, at Madison Square Garden. The engine was tended by Anton Böttcher, a young engineer employed by Diesel and sent over from Germany for this purpose. It was hooked up to a direct-current dynamo and, according to Colonel Meier, ran very well, producing hardly any fluctuation in electric current. This was actually the first exhibition of a diesel engine to the public.[33]

In the summer of 1898, diesel engines were included as part of the Second Power Machine Exhibition, in Munich. This took place on the so-called Kohleninsel, in the Isar River, site of the present-day Deutsches Museum. A separate pavilion was constructed for the diesels. Four companies took part: Augsburg, Krupp, Nurem-

Diesel engines at the Second Power
Machine Exhibition, in Munich, 1898.
(Special Collections Department, Deutsches Museum)

berg, and Deutz. The engines varied from 20 to 40 hp and ran water pumps as well as a Linde air-liquification machine. Diesel's enemies were also there, trying to find something wrong with the engines. Although a number of difficulties were encountered, including very noisy starting, by and large the exhibition was a huge success, with numerous dignitaries, such as Bavarian Prince Regent Luitpold, in attendance.[34]

Although the exhibition proved to be a kind of high watermark for Diesel, he was unable to spend much time there on account of his poor health. His continuous exertions and overwork were leading rapidly toward a breakdown. His description to his wife of his exhaustion after contract negotiations has already been noted. One of his collaborators during this period, Ludwig Noé, described their first meeting in the spring of 1898. Diesel complained of headaches, insomnia, and being so nervous he could hardly marshal his thoughts. Diesel told Noé:

> The development of his engine caused him great worry; he had many enemies and it was hard to defend himself because of his health. In addition, his engine was not yet ready for sale, various parts had to be fundamentally altered before one could view the diesel engine as foolproof. . . . On the way to the train station, Herr Diesel repeated again and again that his nerves were shot, and he had to go immediately to the mountains.

After the visit of the prince regent, "Herr Diesel [again] told me his health was completely at an end, and he had to go immediately into a sanitarium in the Tyrol."[35] Finally, in the fall of 1898, he entered a sanitarium near Munich. In January his doctors sent him to Meran, in the Tyrol, where he remained until April 1899. For these reasons, he was incapacitated during a critical six-month period in the history of his engine.[36]

### Attacks on Diesel's Patents

One of the major problems facing Diesel in late 1897 and in 1898 was the threat of legal action against his main patent. According to German law, an invalidation proceeding could only be initiated within the first five years after a patent had been published; otherwise, it was incontestable. Because Diesel's patent had been published in February 1893, proceedings could be begun only until February 1898. Because the Kassel lecture in June 1897 constituted the first public announcement that his engine appeared to be a success, it was not surprising that anyone wishing to overturn the patent would now come forward before the February 1898 deadline.[37] The two challenges that materialized were made by Otto Köhler and Emil Capitaine.

It has already been noted that Köhler's 1887 book, *Theorie der Gasmotoren*, anticipated Diesel's ideas quite closely. Further, his criticism of the workability of Diesel's theory was on the mark and was probably one of the reasons Diesel made necessary corrections. Diesel, however, never publicly answered Köhler.

In 1897 Köhler appeared on the scene again as a technical consultant for Deutz. He prepared the draft of an invalidation proceeding for the firm, and a copy of his book was sent along with this draft to the Augsburg Engine Works. Deutz apparently decided not to clash with the Augsburg-Krupp combine and did not pursue its attack, but, after the Kassel meeting, Köhler appears to have again threatened a proceeding. At the same time, he apparently suggested a compromise settlement. As early as March, Krupp had expressed fears that the Diesel patent might be invalidated if Köhler chose to attack it legally.[38] Because much of what transpired concerning this matter involved oral conversations among Köhler, Diesel, and Krupp, the situation has never been fully clarified. Suffice it to say that Krupp approached Köhler and offered him 3,000 marks yearly if he would agree to stop opposing the diesel interests and join the consortium. He happily accepted this proposal in early August 1897, remained loyal to the agreement, and never again threatened Diesel.[39]

Just as this agreement was being signed, another threat appeared in the form of a legal suit filed by the Frankfurt engineer Emil Capitaine on July 31, 1897. Since the mid-1880s, he had been developing several oil engines that used a heated vaporizer for ignition. Such engines are looked upon today as an interim step between gas engines and the diesel engine.[40] Capitaine's engine used part of the incoming air to vaporize the fuel. Then the air-fuel mixture was joined with the rest of the air, compressed and "driven up against the redhot walls of the vaporizer, and fired."[41] Capitaine's engines worked with 16 atms compression, considerably more than had been used in previous engines. The engines were built by several companies, chiefly by the Swiderski Engine Works, in Leipzig. In 1891 Capitaine had taken out two patents. One was for a pump that provided air to vaporize the fuel charge. The other was for a device that aided in vaporizing the fuel. Capitaine claimed that his patents anticipated Diesel's main patent.[42]

The legal battle with Capitaine lasted a year and grew more bitter as time passed. It helped to intensify the strain on Diesel, who by April 1898 had become ill. Surprisingly enough, working with his patent lawyer, F. C. Glaser, he apparently attempted to meet Capitaine's challenge alone. Was he afraid his industrial allies would become convinced of Capitaine's claims? Not until the middle of April, after the onset of Diesel's illness, did his lawyer suggest he seek help from Augsburg and Krupp. Diesel accepted the suggestion.[43]

After Diesel had rebutted Capitaine's attack in late September 1897, the latter made two more statements that were also responded to by Diesel. In April 1898 Capitaine made a particularly sharp attack on Diesel in a speech to the Frankfurt VDI that was eventually published. Only one day later, on April 21, the Patent Office in Leipzig rejected Capitaine's suit.

Essentially, Capitaine's attack contained two arguments. He pointed out, quite correctly, that the diesel engine did not incorporate the process protected in Diesel's patent. As he said:

> There are countless cases in the history of invention where the meditating intellect proceeds from and builds on erroneous presuppositions in order to reach a specific goal. The final result of his creative work reaches a completely different goal than that on which he first set his sights. When an inventor openly admits this, his honor and real contribution suffer no damage. It appears dubious, however, when the inventor incorrectly maintains against indisputable facts that he realized exactly what his reflections once told him was the most correct procedure.[44]

Capitaine might well have used this argument as his main line of attack instead of claiming that his patents had already anticipated the diesel process (combustion without significant increase in pressure) and that his engine in fact worked on this principle.

In rejecting Capitaine's suit, the Patent Office held that the originality of the diesel process was characterized by high compression of air and the gradual introduction of fuel into it, followed by isothermal combustion. The Patent Office perhaps thought that this process had actually been realized in the working engine; it did not address this question directly. It further stated that Capitaine's patents did not describe such a process nor did his engines incorporate it.[45]

On July 4 Capitaine appealed the decision to the Federal Supreme Court. He argued that the idea of high compression of air and gradual injection of fuel was well known before Diesel.[46]

At the same time as his appeal, Capitaine approached Augsburg suggesting a settlement of 25,000 marks to end the controversy. He was encouraged to do this by August Klumpp, a patent attorney who acted as intermediary.[47] When informed of this offer, Krupp answered that it was unsympathetic to such a settlement, but was ready to pay a third of the required funds.[48] By July 12 Capitaine had signed an agreement to abandon all legal action against Diesel and to cease further attacks on him and his patents in return for a payment of 20,000 marks.[49] Capitaine, for his part, was probably happy with this arrangement; if he lost again, his professional reputation would be seriously damaged. Although it is quite possible that he would have lost his appeal, his continued agitation against Diesel, especially with regard to the discrepancy between patent and engine, led the Diesel consortium to seek an end to the controversy. In fact, Capitaine maintained his hostile attitude toward Diesel until the former's death in 1907. Diesel's main patent, however, was now safe, protected by the patent court's assumption that Diesel's engine worked according to the cycle described in his patent and the misemphasis of Capitaine's attack.[50]

## Difficulties with the First Engines

During this same period of Diesel's declining health and Capitaine's legal challenge, it became increasingly evident that the diesel engine was not a readily marketable product as almost everyone had thought in June 1897. It would run smoothly on the Augsburg test stand, carefully tended by trained personnel and not subjected to rough handling or undue loads. Once, however, the engine was put into industrial operation, the difficulties began.

The problems were especially apparent in the very first diesel engine sold for industrial purposes. Purchased by a match factory in Kempten, Bavaria, whose director, Karl Buz, was Heinrich Buz's brother, it was constructed in Augsburg in the fall of 1897 and delivered the following March. It was a two-cylinder, 60-hp engine. Augsburg had to agree to replace it with a steam engine free of charge if it proved to be unsatisfactory. When the engine was subjected to heavy loads, major problems were encountered. Augsburg had to send two mechanics and finally Immanuel Lauster to Kempten to tend the engine night and day. Not until a year after its installation, including a complete rebuilding, did it run properly. The director of the Kempten factory is supposed to have said, "The diesel engine cost me ten years of my life."[51]

Industrial use disclosed four basic problem areas. First, the high pressures and temperatures of the engine caused parts to wear out rapidly. For example, pistons corroded, warped, or ate into the cylinder; valves became stuck; oil lines ruptured; and exhaust gases leaked into the engine room. New metals and parts had to be continually tested to overcome these problems.

Second, the single-stage air compressor used for air-blast fuel injection was not only unsatisfactory but quite dangerous. Because the air was compressed in one stage to 50 or 60 atms, it became so hot that it often ignited the lubrication oil in the pump, either clogging the pump or causing explosions. Colonel E. D. Meier's nephew was killed in such an explosion. The problem was not solved until several companies, such as L. A. Riedinger, in Augsburg, and

First diesel engine in industrial use, 1898.
The engine was sold to a Kempten, Bavaria, match factory, run by
Heinrich von Buz's brother. The "childhood diseases" were readily apparent,
and only after a year of constant tending
and a complete rebuilding did the engine run properly.
(M.A.N. Werkarchiv)

the Augsburg Engine Works introduced a two-stage compressor in 1899–1900, which allowed for compressed air at a lower temperature. Ultimately, in the twentieth century, direct fuel injection would replace compressed-air fuel injection.

Third, the sieve vaporizer proved to be unworkable. So long as trained Augsburg engineers were on the job, it could be dismantled at the end of the day and cleaned every night. This involved careful unwinding, cleaning, and rewinding of the brass gauze between the vaporizer plates. If this was not done properly, the engine lost power or even died. Not until the fall of 1899, after much random experimentation, did Burmeister and Wain, in Copenhagen, and M.A.N., largely through the work of Lauster, devise the so-called plate vaporizer, in which the compressed air forced fuel through a series of four plates that contained a number of staggered holes at various angles and vaporized the fuel. Although still a vulnerable point, the plate vaporizer was an essential step on the road to making the diesel engine a marketable product.

Fourth, the diesel engine was quite sensitive to varying types of fuel, which were often changed rather haphazardly during this period and produced significant variations in performance. Although tests were under way on a large variety of fuels at that time, the fuel question was not adequately solved until the twentieth century.

Striking evidence of this problem is provided by Eduard Blümel, an engineer who was hired in 1899 by the short-lived Diesel Engine Company of Augsburg, which is discussed in the next section. In a letter to Eugen Diesel, Blümel described another engineer in the company whose sole job was to build one fuel pump after another. Blümel constructed a functioning fuel nozzle, but, as soon as the fuel was changed, it would fail. Crude oil, sometimes thin, sometimes "thick as syrup," arrived in a variety of containers. The color varied from golden yellow to dark green. One of the engineers responsible for constructing diesels supposedly came to Blümel with a sample of oil and exclaimed: "And we are supposed to work with this muck (*Dreck*)!" Although Blümel's remembrances date from 1940 and he may have been looking back with some hindsight, he seemingly already had a feeling at the turn of the century for the diesel engine's sensitivity to varying fuels, a fact not well appreciated at that time. As he said, "The relationship between the diesel engine and its fuel is like the relationship between the modern high speed locomotive and its roadbed or the automobile and the expressway. If the prerequisite is lacking the enterprise is a failure."[52]

If Augsburg carefully maintained its first commercially sold engines, all companies could not be expected to monitor theirs so closely. By late 1899 firms such as Krupp and Deutz had given up production of diesel engines, foreign companies were in difficulty, licenses were being dropped, and Diesel's once promising business seemed to be on the verge of disaster. Only Augsburg and the Nobel Engine Works, in St. Petersburg, continued their production of engines uninterrupted. The determined work of Augsburg engineers, such as Lauster, finally overcame the difficulties and saved the engine. Buz's unshakable faith kept the engine alive.[53] As Friedrich Sass points out, none of these difficulties involved the diesel cycle. They were all connected with the practical problems of engine construction and the lack of the proper technology and experience.[54]

From 1897 on, Diesel did little to correct the problems besetting the diesel engine and was not consulted by Augsburg concerning them. His efforts were now directed toward exploiting his foreign patents for as much money as possible. Although at various times he directed the Augsburg testing station between 1897 and 1899, working on a compound engine as well as with a variety of fuels and engine improvements, none of these experiments produced results, and Buz closed the station at the end of 1899. The experiments were either unworkable or ran too far ahead of available technology. Buz realized that the main task was to work with the oil engine at hand to turn it into a marketable product. Diesel's creative period was essentially over; his relationship—or lack of it—to further improvements made by Augsburg did not essentially change until he died in 1913.[55]

## The Diesel Engine Company of Augsburg

During this critical period, Diesel became involved in the founding of a new factory in Augsburg, devoted (at first) solely to the production of diesel engines. It was known as the Diesel Engine Company of Augsburg (Dieselmotorenfabrik Act.-Ges. Augsburg) and was organized between the fall of 1897 and the following spring.

The exact background of this enterprise is not completely clear. It is not surprising, however, that, in view of the extremely favorable publicity the engine was receiving in 1897, especially after the VDI meeting in Kassel, interested parties would approach Diesel and try to become involved with the engine in hopes of making a profit. This was true in Augsburg, where in the fall of 1897 two banks—P. C. Bonnet and August Gerstle—approached him through his cousin, Christian Diesel, who owned a transport company in the city.[56] The original idea was to gain control of Diesel's Austro-Hungarian, English, or American patents. He indicated this was no longer possible, but suggested that some other connection with his engine could be worked out.[57] Out of this beginning came the new factory.

By early November, Christian was arranging meetings between the bankers and Diesel as well as reporting on an old factory building that was being considered as the site of the new establishment, and Gerstle was in Munich negotiating with Diesel.[58] On November 13 Gerstle signed an agreement for the consortium with the Augsburg Engine Works and Krupp, which granted an option on a license to build diesel engines.[59] Then, on November 18, Diesel notified all parties concerned that he would not take part in the founding of the new factory because the price of the license (200,000 marks) and the patent royalties required by Augsburg and Krupp (20 percent on engines up to 16 hp and 30 percent on engines over 16 hp) were so high that the prospective factory's hopes for profit were too uncertain.[60] A copy of this letter was also sent to Heinrich Buz. Perhaps Diesel suddenly realized how high the royalties were, now that he was studying the possibility of becoming involved in the construction and selling of engines.

To what extent was Buz involved in these negotiations? Although, as will be pointed out later in this chapter, after the collapse of the Diesel Engine Company, he was anxious to blame Diesel for most of the difficulties and sought to distance himself from the enterprise, Buz's reply to Diesel of November 19 reveals that the two men thoroughly discussed the new company and that Diesel had even suggested Buz's son-in-law, Lucian Vogel, for the directorship. Further, the Bonnet Bank was the principal lending bank for the Augsburg Engine Works. Max Schwarz, a member of the board of directors of the Engine Works, and his brother Carl were both bankers at Bonnet. Carl was to become chairman of the new Diesel Engine Company's board of directors. Presumably, it was either Max or Carl who first used cousin Christian to contact Diesel.[61] On the other hand, neither Augsburg nor Buz held any stock or official position in the company.

It is not surprising that Buz was encouraging the founding of a new company. Because of the extremely high royalties, he stood to gain more from licensing another factory to construct engines than he probably could have made by constructing engines himself.[62] He could not at this time have foreseen the difficulties this new company would create for the diesel cause.

Buz was extremely agitated over Diesel's sudden refusal to be associated with the company. He complained that Diesel was making it look as if he, Buz, was to blame for the exorbitant licensing fees, when Diesel had agreed to them and never before raised any objections:

> The day before yesterday we discussed among other things the question of the founding thoroughly and you didn't have a single objection at that time. I must say that this letter made an indescribable impression on me—I did not believe my eyes and only after reading it twice did I regain my composure.

Such letters discredit the engine business here. . . . You have been very much involved with the founding, that's known all over the city. . . . You should have mentioned these difficulties before. . . . Nothing in the situation has changed in the last months . . . there will be a good deal of discontent from many sides now . . . and in the coming stockholders' meeting I shall have to explain why the inventor himself says I have made it impossible to set up a new diesel engine factory. And this is happening just when we are trying to increase our capital, specifically so we can build more diesel engines!

After I have done so much for the cause, I am really not prepared for such a situation.[63]

Diesel hastily replied that Buz's business arrangements did not detract from his reputation and that he (Diesel) had raised doubts about the high fees before. For other large companies that manufactured diverse products, a risk was justified, but for a small, new company, manufacturing only diesel engines, the risk would be too great. Diesel suggested they meet to iron out their difficulties.[64]

Despite Diesel's reservations, within a month he had reversed himself and signed an agreement with Bonnet and Gerstle on December 15 that gave each party a third ownership in the factory. The available evidence does not explain why Diesel changed his mind. Did Buz's pressure help to convince him? Had he only written the letter of November 18 in order to put pressure on Buz to lower the fees for the Diesel Engine Company, which he really intended to join? Was he convinced by the Augsburg bankers or by cousin Christian, who, whatever his original role in the matter, stood to gain by being named to the board of directors?[65] That Diesel was pursuing his own interests apart from Buz can be inferred from an undated letter to his wife about this time in which he said, "the Diesel Engine Company has come into existence in grand style. . . . Herr Buz will be surprised when a factory which is both solid and has good prospects is set up in Augsburg next to the Augsburg Engine Works. I only fear that Herr Buz will be somewhat envious. So much the worse for him."[66]

Diesel may well have had his reasons for supporting the enterprise. The Augsburg Engine Works had already announced that for the time being it could devote only part of its operation to diesel engines, and had pushed Krupp into seeking Deutz as a licensee. Further, some of Diesel's contemporaries suggested that he either did not wholeheartedly trust Augsburg to develop his engine with vigor or that some opposition may have existed in the steam-engine division to rapid development of the diesel engine.[67] As was pointed out in chapter 4, Josef Krumper was placed in charge of the diesel construction bureau in April 1897, and Diesel did not view him as a friend of his cause. Whether or not Diesel's suspicions were correct, a new company might have seemed a good way to make progress with his engine, as well as a way to make more money for himself. When he was approached in January 1898 by a party that wished him to sell some of his stock in the new company, he refused on grounds of principle and "all the more because the stock represents a very advantageous capital investment for me."[68] Unfortunately for him, his declining health and the Capitaine lawsuit, plus the large amount of business concerning his other patents, meant he could provide little personal supervision over the company. Further, he was not a good businessman and was a poor judge of those he thought might help his cause.[69]

The empty factory site discussed in the letters of November 1897 was purchased for 400,000 marks, and the new company was registered at Augsburg in January 1898. The first business year was to run through September 30, 1899. The licensing contract with the Augsburg Engine Works and Krupp was signed on April 15, 1898. The founders were listed as Carl Schwarz (presumably for Bonnet); Diesel; August Gerstle; Robert Jansen, an Augsburg manufacturer; and Christian Diesel. The company was capitalized at 1,200,000 marks, distributed to the founders equally in a total of 1,200 stock certificates, each of which was valued at 1,000 marks. By prior arrangement, half the stock was then sold on the market. A heavy demand ensued, especially among local Augsburgers. The founders made up the board of directors, which was chaired by Schwarz.

The technical director of the company was Max Behrisch, an engineer from Würzen, and the business director was Emil Krüger.[70] Although the license-purchasing fee was reduced to 100,000 marks, the royalties remained the same as for other companies: 20 percent for engines up to 16 hp and 30 percent for engines of more than 16 hp. Diesel, therefore, did not obtain the lower royalties he might have hoped for. In January 1900 the factory employed 152 workers.[71]

At first, business must have seemed quite satisfactory because a number of orders for engines came in from Germany and from as far away as Penza, in Russia. In September 1898 the Diesel Engine Company sent a list to Diesel indicating that eight engines had already been ordered, ranging from 15 to 50 hp and destined to run electric dynamos, water pumps, and refrigeration machines in breweries.[72] By 1899 the first engines were being delivered, and the company's difficulties began. As one of Diesel's colleagues, Karl Dieterichs, said, "The drama began immediately with what so far as I know was the first [sic] engine, produced for Perm in the Urals. It left the factory on a laurel covered wagon, and a few months later it rolled back through the doors without song or dance."[73]

One source of trouble was the typical "childhood diseases" that beset all early diesel engines, especially those concerning the single-stage air compressor and sieve vaporizer. Such difficulties occurred immediately with what in fact was the first motor, delivered to the Oderbruch Brewery, in Seelow, near Frankfurt.[74]

An interesting letter from Albert Johanning, director of the General Society for Diesel Engines (Allgemeine Gesellschaft für Dieselmotoren), to Diesel discussed the Seelow situation.[75] As Diesel's legal successor, the General Society was naturally worried that the Diesel Engine Company's difficulties might seriously damage diesel engine interests everywhere. The company had assured Johanning that the chief source of difficulty with Seelow lay in the high petroleum prices, attributable to the high tariffs on oil. Although high fuel prices were an impeding factor, as can be seen in other correspondence of the time, the company was clearly having other problems. It suggested that a monetary compensation to the brewery would prevent return of the engine. The engine was 40 hp and had cost 18,900 marks, including a 30-percent royalty of 5,670 marks. Johanning, therefore, suggested that Augsburg, Krupp, and Diesel each give up their third of the royalty, or about 2,000 marks each, and that the money be used for compensation. The General Society's attempts to save the situation, however, failed. In the fall of 1899, the engine had to be returned. Several witnesses testified to the lack of technical expertise and to the fact that the engines were "miserably constructed" at the Diesel Engine Company.[76]

Further problems are described in a Johanning memo of May 8, 1899. Not only were the managers inept, but they also spent much time publicly talking pessimistically, even derogatorily, about their own products, instead of putting a good face on the situation. They also refused to abide by the solidarity clause, in which improvements were to be shared among Diesel licensees. The General Society had sent one of its own engineers, Ludwig Noé, to help with construction problems, but, after several days, he had been politely shown the door. Under the circumstances, the company's stock had dropped in value, and the first year's business report was bound to hurt the General Society. Johanning's suggestion was that the society should take over the Diesel Engine Company, but this idea came to naught.[77]

In September, Johanning appealed to Heinrich Buz to send one of his engineers to the factory to help out in construction. Any more defective engines would cause as much damage to diesel interests abroad as had already occurred at home.[78] Once again, on October 25, 1899, Johanning wrote Buz in the latter's capacity as chairman of the General Society's board of directors. He reminded him that the society held 100,000 marks of stock in the Diesel Engine Company and suggested that he speak to Max Schwarz and urge him to convince his brother Carl to accept a technical adviser.[79]

In September 1899 the Diesel Engine Company asked for and received a reduction in royalties and an extension in the deadline for their payment.[80] In December, Diesel approached Johanning, indicating that the first year's business report would look bad and blaming the deficit on high oil tariffs. He suggested that the General Society, the Augsburg Engine Works, and Krupp temporarily give back their shares of the 100,000 marks licensing fee so that the Diesel Engine Company would not show a deficit on its books.[81] Johanning's answer rather bluntly stated that all of Diesel's excuses were pointless and that, when the public found out about the deficit, they would take out their anger on Diesel. Rather than canceling the licensing fee, Johanning and Buz expected Diesel himself to make up the deficit, perhaps out of profits from an oil consortium he expected to found the following year.[82]

By 1900 all efforts to save the Diesel Engine Company had been fruitless. A license to build Riedler express water pumps had been acquired, but little progress had been made in building or selling them. At a general stockholders' meeting on May 14, the directors were forced to report that the company had failed and the production of diesel engines had to be abandoned.[83] A General Society report stated the various reasons for the company's demise that had surfaced at the stockholders' meeting—high oil prices, poor directors, the "childhood diseases" of the engine—and reported that sentiment existed that the founders should reimburse the stockholders. Diesel's reputation, the report concluded, was bound to suffer from all of this.[84] Apparently, the company remained in existence until 1906, when it went into liquidation. It officially ended in 1911.

After the company failed, the postmortems began, along with attempts to fix blame. In a letter to the General Society, Diesel criticized the directors and poor conditions at the factory. He pointed out that he was ill during the period when the first engines were being constructed and complained that false rumors were now circulating about the diesel cycle itself, whereas the problem lay completely in faulty construction.[85] At other times, he complained that he had been talked into participating in the company by Augsburg bankers.[86]

For its part, M.A.N. hastened to distance itself from the failed enterprise. In a letter that was sent to the *Frankfurter Zeitung* and which was reprinted in the *Augsburger Zeitung*, M.A.N. claimed to have no closer relationship to the Diesel Engine Company than to any other licensee, to have taken no part in its founding, and to hold no stock. In addition, M.A.N. claimed all the "childhood diseases" had been overcome and that the only reason more engines were going to Russia than to Germany was the high German oil tariffs.[87] Although some of this disclaimer was true, it has already been noted that Heinrich Buz was involved in the negotiations leading to the establishment of the factory. If by 1900, the "childhood diseases" had been mostly overcome, this had not been true a year earlier.[88]

According to Immanuel Lauster, Buz was extremely upset with Diesel and at one of the meetings of the General Society's board of directors indicated that he held him personally responsible for the failure of the factory.[89] In a letter to Diesel, Buz complained:

> So long as such miserably constructed engines as those in Breslau, Pensa, and Arad, etc., etc., are not removed from this world, they will remain a continual drag on the diesel cause, and I deplore the fact that our competition has been given ammunition with which to successfully attack us.
>
> Do not take me amiss when I allow myself the observation that, if I were in your place, I would exchange all these engines at my own cost. The damage has been done exclusively by the Diesel Engine Company, and to my knowledge your loss can to a large extent be covered by the royalties which you received at that time. Beyond this, your income from the diesel engine, compared to our losses, which are still considerable, is sufficient so that such a sacrifice should not be avoided. Such losses are only temporary and with time will turn into profit.

Diesel answered: "Measures for the definitive removal of the unsatisfactory engines are underway, and I hope that a conclusion to the affair, agreeable to all parties, can take place."[90]

What precisely Diesel did is unknown, but it is obvious that the collapse of the enterprise represented a financial blow for the General Society, and therefore for him, as well as being a blow to his reputation.

As was to be expected, the stockholders were furious that their money was lost. Scapegoats were sought. The *Augsburger Stadts-zeitung* of October 4, 1900, carried an article entitled "Who is to Blame?" that called the whole affair "a great scandal." How, it asked, with M.A.N. producing functioning engines and Diesel personally behind the company, could it have failed in only two years? Obviously, the stockholders had been cheated, while the bankers and Diesel continued to prosper, and the diesel engine won the grand prix at the Paris World's Fair. The paper even carried a sarcastic poem on the whole affair about a stockholder who supposedly committed suicide.[91] (Of course, one ultimately would: Diesel himself).

Paul Meyer, who visited the Diesel Engine Company factory, expressed the opinion that, though the directors were inept and the conditions bad, the real problem was the difficulties that beset all diesel engine companies at the time. If M.A.N. was having difficulty, how could a small factory, where the personnel were inexperienced and the equipment old, hope to succeed? The factory was a victim of the generally held but mistaken view that the diesel engine was a fully marketable product in 1897.[92]

Added to these problems were poor management, Diesel's incapacity at a critical time, and the apparent inability of M.A.N. and the Diesel Engine Company to work together. Some evidence even exists that Diesel told the company not to cooperate with M.A.N.[93]

Although Diesel had only been slightly involved with the technical side of the company, popular belief held that he actually directed it and was, therefore, responsible for its failure. This view was espoused by some of his critics as late as 1912.[94] He lost the money he had put into the company's stock, and there is some indication he paid back part of its debts.[95] This was the beginning of his financial ruin.

The reputation of the diesel engine suffered a severe blow, from which it only slowly recovered. According to one account, the flood of pro-Diesel newspaper publicity that had occurred after the Kassel meeting was reduced to zero by this "fiasco."[96] Paul Meyer reported that he left the employment of the General Society for Diesel Engines in the middle of 1900 because diesel production was at its nadir and because at that time no hope of employment existed elsewhere in the diesel business.[97] Lauster stated that he was told on a business trip to Russia that the Diesel Engine Company's failure was the main reason why firms hesitated to order engines. Furthermore, a number of people confused the company with the Augsburg Engine Works and thought it was the latter firm that had failed.[98]

Supposedly a bank wanted to cut off credit to a customer who bought a diesel engine, and an electric firm refused to send dynamos to installations that were going to power them with diesel engines.[99] The whole affair was the beginning of Diesel's slow estrangement from Heinrich Buz, the man who more than anyone else besides the inventor was responsible for the development of the engine. Thus, a combination of Diesel's own failings and an erroneous view of the marketability of the engine created a near disaster for the diesel cause.

### The General Society for Diesel Engines

One of the most fateful enterprises that Diesel involved himself in during the late 1890s was the General Society for Diesel Engines (*Allgemeine Gesellschaft für Dieselmotoren*), set up in September 1898. The impetus for forming it came from him, though it is not known how long he might have been contemplating such a move or to what extent he may have discussed it beforehand with

business and banking interests. In any event, during the Second
Power Machine Exhibition, in Munich, in the summer of 1898, he
wrote to Berthold Bing, now head of the Russian Diesel Engine
Company, in Nuremberg, proposing that a company be founded
that would take over all of Diesel's stocks, royalties, and rights to
patents as well as licenses both present and future in return for a
one-time compensation of three and a half million marks. In essence, Diesel was proposing to consolidate all his business affairs in
one company and to give up direct control over his patents and
engine, that is, over further direct financial exploitation of his invention, in return for a large financial settlement.[100]
  Diesel gave his reasons for taking this step in the letter to Bing:

> Because of the enormous growth of my business, which at the mo-
> ment has extended so widely that I cannot control it and which
> threatens to overwhelm me, and because of my latest nervous ill-
> ness, which has given rise to the worst fears, I have gotten the idea
> that it is necessary to place my affairs, which are now doing quite
> well, on a secure and broad basis and to make them independent of
> my person.[101]

Diesel pointed out in both this letter and in a subsequent one to
Heinrich Buz that he wished to be rid of his business concerns so
that he could devote himself again to the "one real goal of my life,
the technical perfection of my engine."[102]
  These reasons seem to make sense. Diesel had made himself the
center of a series of international licensing agreements, based on
the reciprocity clauses. By 1898, however, he could no longer hold
all the threads of his business enterprises together. Further, as he
said to his wife, he was especially anxious to provide financial security for himself and his family, independent of his own fate or the
further progress of the engine.[103] The summer of 1898 seemed the
right time to act; it marked the high point of his affairs. Already,
however, difficulties with the first engines were becoming apparent. Waiting any longer might produce much less favorable business conditions.

Diesel argued that he was not particularly fond of banks and that the companies that had had most to do with developing the engine, the Augsburg Engine Works and Krupp, should be the ones to reap the benefits of the enterprise. By doing so, the royalties they were supposed to pay would actually "flow back to their own pockets." He convinced Bing to take over the negotiations to found the enterprise. Already in a memorandum of June 30, 1898, Diesel had summarized his business dealings and concluded with figures that showed he possessed some 830,000 marks worth of stock in diesel enterprises that provided a guaranteed annual income of 90,000 marks, which was expected to go higher.[104]

Bing's efforts succeeded. The Augsburg Engine Works was willing to join and so was Krupp, with some hesitation.[105] After pressure was applied by Diesel to hasten the matter, the General Society for Diesel Engines was founded in Augsburg on September 17, 1898. Preferred stock in the amount of 1½ million marks and common stock in the amount of 2 million marks were issued, for a total capital of 3½ million marks. In addition to Diesel, some of the holders of preferred stock were, for example, Berthold Bing; Buz; Krupp; P. C. Bonnet; the Augsburg Engine Works; A. Rieppel, in Nuremberg; Markus Wallenberg, in Sweden; Emanuel Nobel, in St. Petersburg; Adolphus Busch, in America; and Sulzer Brothers, in Switzerland.[106] At first, all the common stock was held by three banks. Diesel received a cash payment of 3½ million marks, 100,000 of which went to pay for the costs of founding the company. A few weeks later, however, he purchased 250,000 marks worth of preferred stock and the entire 2,000,000 marks of common stock. This purchase was apparently part of the original settlement, agreed to by him in order to win the acceptance of the other parties involved and to minimize bank participation.[107] It may have reflected his optimism in the new company's future, but it was a major mistake because the stock proved to be worth nowhere near its face value. Heinrich Buz was chairman of the board of directors; Diesel and Max Schwarz, among others, held seats on the board. Thus, from the society's inception, a close connection existed with the Augsburg Engine Works, one that would tighten over time.[108] The director of the company was Albert N. P. Johanning, a Munich businessman.

A key part of the sales contract between Diesel and the General Society was the reciprocity, or solidarity, clause, which stipulated that all parties to the company, including Diesel, were bound to share improvements with one another. Such clauses had in one form or another been written into all his earlier contracts, and he placed great hopes on them leading to rapid advancement in his engine enterprises. On September 17, 1898, circulars from Diesel and the General Society were published announcing that he had assigned his patent and licensing rights to the society.[109]

Although the society was not involved in building or selling diesel engines, it did have a technical office attached to it until 1901 and considered part of its business to be advertising the advantages of the engines as well as trying to coordinate and further various associated enterprises. The numerous reports, letters, and memos of the indefatigable Johanning reveal the society's chief areas of involvement. Although he was always ready with a recommendation for improvement, much of what he suggested was not acted upon.

The society's growing concern over the disastrous history of the Diesel Engine Company in Augsburg was pointed out earlier. The generally poor results with the early engines created a bad press, something the General Society had to combat. In October 1900 Johanning argued in a memorandum that it should involve itself forcefully in countering antidiesel sentiments and propaganda, and also suggested that the society become a central sales bureau for Germany.[110] The latter proposal, however, was never accepted. Various brochures and boards of directors' reports stressed such items as the validity of the patents, the reliability of the engine, its superiority over the steam and otto engine, and the value of the reciprocity agreements. Much of the blame for slow sales was attributed to high oil tariffs and fuel prices.[111]

In a propaganda tract issued late in 1901 entitled *Der Diesel-*

*motor, seine Entwicklung und volkswirtschaftliche Bedeutung,*[112] Johanning blamed the diesel engine's faults on production difficulties in certain factories, such as the Diesel Engine Company of Augsburg, and not on the cycle itself. Interestingly enough, he still placed special emphasis on the engine's role in small industry, which would lead to industrial decentralization. The engine was portrayed as the cleanest and safest one available. A further claim was that it would open up the oil rich areas of the world to industrialization as well as solve the problem of rapidly disappearing coal supplies. It could be used in everything from ships to agriculture and the electrical industry.

Special attempts were made to interest electrical companies. In March 1901 Diesel approached the Allgemeine Elektrizitäts-Gesellschaft (German General Electric Company) (A.E.G.) on behalf of the society and stressed that the diesel engine would open the petroleum-rich areas of the world to industrialization, and thus to electrification. The engine was especially suited for driving electrical dynamos. Diesel did not forget to emphasize the financial gains that would be realized in all of this.[113] Out of the ensuing conversations came an agreement that the A.E.G. would call special attention among its customers to the use of diesel engines with electrical dynamos.[114]

In the General Society documents and various promotional brochures published around the turn of the century, the chief sales pitch in all cases was fuel economy. The diesel engine's thermal efficiency was portrayed as between 28 and 30 percent, while the best steam engine was gauged at 12 percent, and the best oil and gas engines at 16 to 19 percent, respectively.[115] Such emphasis was necessary because the price of a diesel engine was usually about double that of a comparable gas engine.[116] Other features of the engine that were emphasized were safety, size, comparative noiselessness, clean exhaust(!), and immediate operational readiness. In addition, its use for small industrial purposes was often stressed. As the Diesel Motor Company of America proclaimed, "No More Smoke, No More Ashes, No More Explosions."[117] The engine was still conceived of as a stationary engine, meant to replace steam or gas engines.[118]

One of the major concerns at that time was the high tariff on oil, which was 7.20 marks per 100 kg in 1905 and placed the diesel engine in a disadvantageous position compared to other engines, especially in Germany and France.[119] Business reports of the General Society of 1901 and 1902 pointed out that more than half of the diesel engines built in Germany were sold abroad, especially in oil-rich Russia.[120]

Diesel had originally hoped his engine would run on powdered coal, desirable in view of coal's cheapness and availability. Even in the late 1890s, he hoped a solution would be found to this problem and urged Buz not to give up hope.[121] However, by 1899 it was the oil diesel engine that had been developed and on which emphasis needed to be placed.

In the spring of 1899, shortly after recovery from his illness, Diesel became involved in the unwise purchase of oil rights in Galicia. He believed that only in such a way could fuel supplies be obtained for his engine. He attempted to interest the General Society in purchasing additional rights. The board of directors studied this proposal as well as the taking over of Diesel's options, but apparently decided not to make such purchases.[122]

Nonetheless, M.A.N. and Krupp addressed several petitions in 1898 and again in 1900 to the German upper house, the Bundesrat, advocating that oil used for fuel in internal combustion engines be declared duty free.[123] Part of the problem was undoubtedly the fact that since 1879 tariffs had been one of the chief sources of income for the German government, which could not as yet levy direct income taxes. Then, too, coal interests in the country may well have desired to keep oil tariffs high. Tariffs on certain types of oil were reduced in 1906 and again in 1912.[124] They were not removed completely until World War I, but were reimposed after the war, so that oil prices were at a higher level than in the oil-producing countries of Russia and the United States.[125]

The General Society's papers provide some interesting information on the total number of diesel engines in use and their horsepower around the turn of the century that is summarized below:[126]

| Date | Total No. of Engines | Total Hp | Average Hp per Engine |
|------|------|------|------|
| February 1899 | 77 | 1869 | 24 |
| June 1901 | 138 | 4200 | 30 |
| September 1901 | 233 | 8146 | 35 |
| June 1902 | 350 | 12000 | 34 |

These figures show that during the most difficult period, from 1899 to 1901, growth was very slow, not even doubling in two years. Starting in 1901, however, production picked up, rising by 60 percent in just three months and 67 percent in another nine months. However, just how small a percentage this was of the total number of power engines in use can be seen from the fact that, already in 1895 in Germany alone, 18,070 internal combustion engines (mostly otto gas engines) whose total hp was nearly 65,000, or an average of 4 hp per engine were in use. That same year in the country some 58,500 steam engines, whose combined hp was 2,700,000, or an average of about 46 hp per engine, were being used.[127]

A comparison of these figures shows that the gas engine, ironically, more nearly satisfied small power engine requirements, while the diesel engine compared in average hp with the steam engine. These reports verify that about half the engines built in Germany were going abroad, mostly to Russia and Austria-Hungary. In 1899 more were being constructed outside of Germany than inside, but by 1901 the figures had reversed themselves.

As of 1901, diesel engines were still overwhelmingly stationary. Most of them were used for driving electric dynamos; for factory purposes, such as running printing presses, water pumps, and grinding and gear machines; or used in leather works, shipyards, mills, match and lumber companies, railroad repair shops, and weapons factories. The engine was just being developed for marine purposes, and its use in trains as well as trucks and automobiles was still in the future. Thus, by 1901, slow, steady progress had been made in production, but the market was still far behind that of other power engines. This would change as the diesel was improved and adapted to new uses during the next several decades.

Despite Diesel's hopes that the creation of the General Society would remove business responsibility from his shoulders and ensure his financial future, his relations with the company led to new grief and financial loss. As the major stockholder and a member of the board of directors, he could not be free from involvement with the society. The failure of the Diesel Engine Company in Augsburg and the inability of such companies as the French Society for Diesel Engines to succeed meant that the society was holding a large amount of virtually worthless stock it had taken over from Diesel. In addition, the slow growth of sales meant that royalty income was far below expectations. Diesel's relations with the General Society showed strains early on, and several letters indicate differences of opinion as to exactly what their mutual contracts had called for.[128]

By 1901 it had become necessary to reduce the society's capital by 500,000 marks to buy back preferred stock from stockholders.[129] Diesel was involved in this transaction and was supposed to contribute some 95,000 marks. In a letter to Johanning, he stated that he could not contribute his share until the end of the year because the expenses for his new Munich villa had put him in a bad financial position.[130]

By the end of 1903, Johanning had left his position as head of the General Society. He was apparently not only upset about the failures of various diesel engine companies to honor their reciprocity clauses, but was also in the process of being sued for receiving stock from the British Diesel Engine Company and withholding that information from the society.[131]

The reductions in the society's capital continued. In 1905 Diesel had to agree to trade his entire 2,000,000 marks worth of common stock for a "deferred share" (*Genussschein*) worth 250,000 marks, a major financial loss for him. In return, he was released from the September 1898 contract except for the necessity of notifying the society of improvements made in his engine. The exact terms of this new agreement were disputed throughout 1905. Finally, Diesel, upset with the dispute and disliking the new director, Emil Guggenheimer, who was also an attorney for M.A.N., resigned from the society's board of directors.[132] Diesel apparently felt at this time that M.A.N. exerted too much influence on the society for its own good.[133] By 1906 he wished to see the society dissolved and was suggesting plans for its undermining to the American Diesel Engine Company.[134]

One of the General Society's main problems throughout this period concerned various companies not honoring the reciprocity clauses in their contracts. What had been conceived as a way to avoid conflict and speed up improvement-sharing turned rapidly into a source of bickering and contention. The society soon found itself in the middle of these disputes, blamed by companies that felt they were being cheated. Much of the trouble was caused by M.A.N.'s increasing unwillingness to share improvements, especially designs, with other companies. Apparently, it was tired of always giving information to small companies and never receiving anything in return.[135] Also, the French Diesel Engine Company, under government pressure, was increasingly refusing to transmit improvements on engines used for military purposes. A letter of October 3, 1902, from Colonel Meier, of the American Diesel Engine Company, complained that the society was violating its contract by refusing to send drawings for a new 8-hp engine developed by M.A.N. The American company threatened to either break the contract or demand a damage payment of 50,000 marks.[136]

In November 1902 another letter came, this time from the Diesel Engine Company, Ltd., in England:

According to our agreements, we have the right to all improvements made by all firms making Diesel Engines, and we cannot even exclude any types being made for the French Admiralty nor can we regard political reasons. . . . you quite forget that you are under engagement to suply [*sic*] drawings to us of every size and every type that is made, also that you have to supply us with all improvements made in respect of the engine by any of the builders of Diesel Engines, whether such improvements are patented or not. We find now that you are only at home to receive, but when it is the question of fulfilling your engagements, you are not at home.

In order to facilitate matters for you, we have been endeavoring to obtain drawings from the Vereinigte Maschinenfabrik at Augsburg [M.A.N.], but they put such difficulties in the way and raise such delay that we are put to considerable losses . . . *it only shows you the spirit in which everything is done, namly* [sic] *every firm looks out for itself; no assistance is given, and obstacles are put in the way. . . . It seems to us that all firms making Diesel Engines simply look after their own interests and do not keep their engagements to you. . . .* There has been too much talk, and to [*sic*] many promises, instead of your engagements being kept, but we are tired of this sort of thing, and shall in future insist upon getting what we bargained for.[137]

The General Society's answer in both cases simply stated that it could not hand over the designs in question because M.A.N. had not made them available. The society could not give what it did not have.[138]

In July 1903 M.A.N. agreed to deliver designs of engines that had been thoroughly tested. In return, it received 15 percent of all income of the General Society, plus export rights to Spain and eventually to Portugal and Italy.[139]

By December 1903 the French Diesel Engine Company had voted officially not to transmit improvements on engines used for military purposes, whereupon the Diesel Engine Company, in Budapest, and Sulzer Brothers, in Switzerland, refused to honor their reciprocity agreements. The General Society sued Sulzer but lost.[140]

Relations between Diesel and both M.A.N. and the General Society were at the breaking point by 1906. The exact reasons for his falling out with M.A.N. are unclear. It may be attributable to his feeling that M.A.N. had not pursued his interests with as much vigor as he would have liked. His mercurial personality and inability to keep out of business affairs probably contributed. Undoubtedly, the Augsburg Diesel Engine Company affair had strained relations. The difficulties created for the General Society by M.A.N.'s refusal to share designs of improvements and its increasing influence over the society may have served as a further irritation. Eugen Diesel casts most of the blame on the General Society's new director, Emil Guggenheimer (though not mentioning him by name). Supposedly, whenever the elder Diesel spoke to his family about the matter, he most often mentioned Guggenheimer but rarely Buz.[141] Diesel had no love for Guggenheimer. On December 20, 1906, he wrote Buz complaining of the way Guggenheimer was running the society. Guggenheimer wrote Buz on December 23 taking sharp exception to the criticism.[142]

In late 1906 Diesel announced to the General Society—which passed on the information to M.A.N.—that he was about to patent a new process that, among other things, reduced the amount of pressurized air needed for fuel injection. Although the application was rejected by the Patent Office, both the General Society and M.A.N. demanded to know further details of the process. When Diesel refused, both companies sued him in February 1907.[143] The suits dragged on until 1909. The M.A.N. suit was rejected on the basis that Diesel's obligation to make improvements known had been taken over by the General Society and that, further, his main patent had expired in February 1907. M.A.N. appealed, but the issue was settled out of court in 1909, when an agreement was reached to split court costs and to have each party pay its own attorneys' fees.

The General Society suit was at first decided in Diesel's favor, but the society won an appeal. Diesel, too, apparently was considering appeal,[144] but by 1908 the society's board of directors was recommending liquidation because Diesel's main patents had run out and most of the firms involved did not wish to see the society continue. In view of the impending liquidation, the society had little interest in pursuing lawsuits against Diesel. Although it had continued to reduce its capital stock, it was able to pay interest of from 5 to 6 percent on the remainder. On February 27, 1911, the society was dissolved; all remaining rights and duties passed to M.A.N. and to Krupp.[145]

Thus, what had begun as a potentially smart business move on Diesel's part led in the end to financial loss; and, instead of solidarity, to multiple lawsuits among Diesel; his main industrial backer, M.A.N., and his own creation, the General Society. The founding of the society was based on overly optimistic ideas of the marketability of the engine and on the solidarity clauses. Diesel's heavy purchase of society stock was also a mistake. Because of his involvement in the organization and his inability to leave well enough alone, tensions were bound to grow. By 1906, with Buz as chairman of the board and Guggenheimer as director and with its headquarters at M.A.N., the General Society was essentially controlled by M.A.N. At this point, relations among the society, M.A.N., and Diesel were so strained that any issue could have touched off a lawsuit. Certainly, Diesel's announced improvements to his engine in 1906 were hardly in themselves worthy of such trouble because they were never patented and proved to be useless. In view of M.A.N.'s earlier refusal to honor reciprocity agreements, it seems ironic that the company would sue Diesel over the very same issue. The British Diesel Engine Company's remark was a sad but true commentary on the fate of Diesel's optimistic hopes: "It only shows you the spirit in which everything is done, namly [sic] every firm looks out for itself; no assistance is given, and obstacles are put in the way."

### The Final Years

Although the diesel engine's fortunes were at a low ebb around the turn of the century, the situation began to improve after 1900.

During the last thirteen years of Diesel's life, slow but steady improvement was made in the engine and in its development for use in a number of fields. Although it never realized Diesel's original high hopes, it found a firm place alongside the otto and the steam engine.

Diesel's main patent expired in February 1907, and his second patent in November 1908. Now numerous firms, such as Deutz, that had either given up construction of the diesel engine around the turn of the century or had never constructed it, hastened to begin production, largely because of the successes of M.A.N. For example, Deutz announced its intention of selling diesel engines again, starting March 1, 1907, one day after the expiration of Diesel's first patent.[146]

In 1908 M.A.N. gave up the manufacture of steam engines altogether in favor of diesels. It was able to arrange a price convention signed by German companies that forbade undercutting of competitors. This convention expired in 1910, and from then on firms had to compete directly against one another on the basis of their engines' prices and advantages.[147]

Diesel took no direct part in the major technical improvements of his engine because his relationship with M.A.N. was increasingly strained, and the General Society controlled his patents. His main contribution during the first decade of this century involved his many trips and several publications on behalf of the engine.[148] During these years, he collaborated to some degree with Sulzer Brothers on a diesel engine locomotive and did some work on the first truly small (4–5 hp) engines, but overall his inventive efforts were in the past; he was never able to repeat his success of 1892–97.[149] By 1912 both his reputation and that of his engine were growing; he was by then quite well known and often feted on his trips. His personal financial situation, however, grew ever more precarious, and his health became progressively worse. He would not seek financial help or advice from wealthy friends, and his involvement with new companies, such as the Busch-Sulzer Brothers

Diesel Engine Company, in America, came too late to change his financial situation.[150]

In November 1912 Diesel delivered his last important lecture, before the German Society of Naval Architects at the Technische Hochschule in Berlin-Charlottenburg. The meeting was chaired by Friedrich August, the grand duke of Oldenburg. After Diesel finished his lecture, the following discussion period was dominated by two professors, Alois Riedler and Adolf Nägel. Launching a combination of attacks on Diesel's personal integrity and inventive originality, they mentioned the money he had made, and especially Nägel hinted that he might have falsified his test journals to play up his role in the engine's development. The audience greeted these personal attacks with open hostility.

More seriously, Nägel accused Diesel of not mentioning his most prominent predecessors, such as Söhnlein and Capitaine. Riedler drew distinctions among inventions that are "working" (*gangbar*), "useful" (*brauchbar*), and "marketable" (*marktfähig*). The last two types were, according to Riedler, the most significant and, in the case of the diesel engine, were the accomplishment of M.A.N., not Diesel. Further, his original idea of realizing isothermal combustion had not been workable anyway, so in a real sense M.A.N., not Diesel, was the creator of the diesel engine. His main accomplishment was the financial exploitation of the engine.[151]

Diesel's answer both at the end of the discussion period and in an appended statement to the published version of his speech emphasized that at the time he conceived of his engine, he was not aware of his major predecessors' work and that in his lecture he had only meant to discuss the period of the engine's origin up to 1897, not its subsequent development. He later appended statements from 1897 to show that the Augsburg Engine Works and Krupp had then judged the engine to be marketable. Diesel also claimed he had on numerous occasions pointed out that his theory had been modified.[152] To his credit, he refused to answer the personal attacks directed at him.

Riedler was wrong in not crediting Diesel for the inventive and initial developmental stages of the engine, though his claim that M.A.N. essentially made the engine marketable was correct. Unfortunately, Diesel himself did not distinguish clearly enough between the periods before and after 1897 and stressed too much the marketability of the engine as of 1897. Thus, although his lecture and the book based on it, *Die Entstehung des Dieselmotors,* helped clarify many aspects of the Diesel story, they were still not a completely satisfactory explanation, and criticism of him continued.

Partly for this reason, Diesel took his own life less than a year after the Berlin lecture. What had seemed such a major victory in 1897 had led some fifteen years later to a personal tragedy that would prevent him from seeing or sharing in the triumph of his invention.

# CONCLUSION

•

In assessing the significance of Rudolf Diesel's life and work, four major areas may be stressed: the role of the inventor and the relationships of engineering to social issues, science to technology, and technology to industry. Examination of these areas will allow some conclusions to be drawn about the nature of the process of invention, development, and innovation and about Diesel's relationship to the industrial revolution at the turn of the century.

A commonplace argument would be that a high-pressure oil engine would have been devised around the turn of the nineteenth and twentieth centuries, even if Diesel had not been involved. The desire to invent a more economical engine was strong, and hundreds of inventors were working to achieve this goal. Yet, the significance of the individual in both the successes and failures of a technological enterprise cannot be overlooked.

Diesel possessed many of the qualities necessary for success: intelligence, drive, tenaciousness, and the ability to make such a good case for his ideas that he carried others along with him. He conceived the idea of realizing a more efficient heat engine than the steam engine from a college lecture. The idea apparently obsessed him and against formidable odds he saw it through to an apparently successful realization by the late 1890s. However, his unstable personality, propensity for overwork followed by collapse, suspicion of his colleagues, underestimation of the technical difficulties involved in the development and innovation phases, and his grand airs and desire to live the life of the successful millionaire-inventor—all contributed to the near-failure of the early engine and to his personal undoing.

Diesel cannot be blamed completely for the failures of his enterprise between 1898 and 1900. His illness kept him out of action during a critical period. Furthermore, almost everyone else made the same mistake as he in thinking the engine was marketable. However, he helped ruin himself through unwise business speculations and the extravagance of his Munich villa. He is a good example of the successful inventor who is a poor businessman.

Diesel also illustrates German engineers' attitude toward social problems. The engineering profession that arose in industrializing countries during the course of the nineteenth century found itself a latecomer among the more established professions, such as law and medicine. Complaints about lack of prestige were common among engineers, but the German situation was made especially difficult by the conservative nature of society there and by the romantic distrust of technology. In Germany, perhaps more than in other countries, engineers turned to an examination of cultural problems and proclaimed technology as the basis of cultural advance.

Diesel shared the mid-to-late nineteenth-century enthusiasm for reason, science, and technology. He was genuinely concerned for the workers and was involved, as were others like Franz Reuleaux, in trying to solve social problems. He, too, subscribed to the industrial decentralization theory and thought of his engine as the small power source destined to stem the tide of proletarianization. He shared the widely held idea that engineers were a neutral third force, mediating between capital and labor. Confident that the most rational solution to a problem must succeed, he turned to outlining a solidaristic restructuring of society, which would solve social problems while leaving the bases of capitalistic society untouched. His solidarism was largely cast in the form given it by his reading of the French utopian socialists. Therefore, he combined the conservative, status-seeking motivations of many nineteenth-century German engineers with an engineering tendency to propose rationally structured utopias that bore little relation to realistic political and social trends. Thus, he demonstrates the fallaciousness of applying simplistic engineering solutions to social problems.

It is interesting that both Diesel's solidaristic ideas and his original theory of a heat engine sprang from a desire to apply rational-logical solutions, in the one case to social problems, in the other to the problem of efficiency in an internal combustion engine. As was the case with his social ideas, his "rational" solution to the problem of the heat engine did not work. Diesel had the time to modify his engine theory to fit a workable engine. He did not have the time or the inclination to do the same with his social ideas. Also, perhaps social problems are not so easy to solve as are technical ones.

The invention of the diesel engine illustrates a growing relationship between science and technology at the end of the nineteenth century. Diesel was an example of a new type of inventor-engineer, scientifically trained at the Munich Technische Hochschule. His invention was inspired by the comments of his teacher Carl Linde on the inefficiency of the steam engine and on the ideal Carnot cycle. Diesel was familiar with the latest theories of thermodynamics, especially through the work of Gustav Zeuner. Although he had some practical experience in Linde's company and in working on his ammonia engine, his approach was basically theoretical, working out the elements of a process that would realize isothermal combustion. He stressed thermal efficiency so much, however, that he neglected to consider other efficiencies. Indeed, Lynwood Bryant has referred to Diesel as being basically an "efficiency nut."

Diesel also seemed little aware of the difficulties of building a working engine. Interestingly enough, leading academic experts, such as Schroeter and Zeuner, also seemed not to recognize the practical difficulties involved. The original doubts came from engine practitioners. Diesel's theoretical boldness led him more rapidly into an area that he might have been slower to enter if he had possessed more practical experience with engines. He was not only forced to modify his theory, but he also encountered numerous practical difficulties and delays during the period of development because his engine all but outran the practical technology of his day.

In the early history of the steam engine, practice outran theory. Indeed, it called forth the study of thermodynamics. By Diesel's time, scientific theory had advanced beyond practical engine building. If science and technology were growing closer together, no neat correlation yet existed between them. Rather, they mutually interacted and influenced each other's development. This kind of dialectical interaction between science and technology is perhaps a truer picture of reality than a scheme that regards technology as applied science.

Especially in the area of invention, the difficult problems of originality and priority are encountered. Priority in invention is often a disputed matter, but in Diesel's case the problem seems especially acute. He did not originate most of the components of his theory. Indeed, his ideas on how to achieve isothermal combustion in a compound engine bear a close, if not exact, similarity to the ideas of Otto Köhler. In addition, a number of other inventors were experimenting with high-pressure engines and approaching the idea of compression ignition. Diesel was, however, the first one who put together a number of ideas, modified them into something workable, obtained industrial support, and oversaw tests that produced a working engine. Clearly, the invention and early development stages of the diesel engine were essentially his doing. This, indeed, must be seen as technological originality: the successful combination of disparate elements into something new—in this case a more efficient high-pressure engine.

The fourth area of interest is the relationship of technology to industry. Because of the complex technology of the diesel engine, Diesel was forced to seek industrial support for the developmental stage. He was extremely fortunate to find a visionary entrepreneurial ally in Heinrich Buz. The two formed a temporarily successful team as had, for example, Otto and Langen before them. Although German industry as a whole and the Augsburg Engine Works in particular were suffering the effects of a recession, Buz had the foresight to test this new system and thereby put his factory at the forefront of a new technology.

In the case of the diesel engine, economic reverses and the drop-off in sales of an older, established technology—the steam engine—helped bring forth a new technology. Further, it appears that Augsburg's steam-engine department in the person of its head, Josef Krumper, did not put major obstacles in Diesel's way, as has been previously thought. Older technologies are, therefore, not always an impediment to a newer technology that threatens their existence.

Also, significantly, Buz and Krupp agreed to back Diesel when neither of their firms possessed much experience with internal combustion engines. Indeed, factories such as Deutz that did were probably quite rightly alienated by the demands of the new technology. The Diesel case shows that a certain amount of naiveté combined with daring may be just as necessary for technological breakthroughs as is past experience in the field.

German patent law proved to be as much a hindrance as a help to Diesel. He needed the protection of a patent in order to obtain industrial support. The Patent Office undoubtedly strengthened his patent draft by combining his numerous claims into two more logically coherent ones. Unfortunately, this meant combining realizable and unrealizable parts of his theory in the patent's first claim. When the working engine deviated from the process described in the patent, he was left without adequate protection for his engine. It may have been luck and the patent officials' unfamiliarity with the development of the engine that saved his patent from being overturned. In any event, fending off the attacks of Köhler and Capitaine worried him and opened him to attack, even after the patent could no longer be legally challenged.

Hindsight readily explains why a number of years passed before the working engine became dependable enough to be truly marketable. Although it may not be surprising that Diesel underestimated the difficulties in producing a marketable engine, it is interesting that almost everyone else made the same mistake in 1897. Part of the reason may have been the desire to quickly recoup the money spent on the initial developmental stage. The main explanation, however, involves the complicated technology of the engine as well as the fact that engineers were not sufficiently familiar with the combustion process and with fuels. A number of years would need to ensue before enough was known about the nature of the combustion process, proper design of the combustion chamber, and diesel fuels to create a reliable engine. Krupp seemed to have had an inkling of both the patent and technological problems that lay ahead when the firm cautioned against hasty marketing in the spring of 1897. Diesel and Augsburg, however, allayed Krupp's fears. Although there are other examples of new technologies that are too hastily marketed before the "bugs" are worked out, in Diesel's case this initial mistake almost proved to be disastrous for his engine.

The moral of this story is that the road from invention to development to innovation, or marketability, is not always easy or straight. Indeed, the process of making a product marketable involves further development, even further invention as refinements continually occur. This ongoing process will be examined in the Epilogue. As Lynwood Bryant has pointed out, the inventive, developmental, and innovative stages often proceed simultaneously and are thus not distinct chronological stages.[1]

Finally, by Diesel's time, the inventor and the engineer were subordinate to industry. It was not Diesel but engineers from companies such as M.A.N. who carried on the developmental and marketing stages after 1897. His health problems explain why he could no longer manage his own business affairs, but his international patent and licensing business had expanded so much that it is doubtful he could have mastered it much longer. The sheer size of his correspondence, much of it written in his own hand, is overwhelming. He represents a phenomenon increasingly evident in the twentieth century: the individual buried under a mountain of paperwork and red tape.

Like Carl Linde before him, Diesel created an organization to carry on his business affairs. Linde, however, had proceeded cautiously with the marketing of his refrigerators.[2] Diesel preferred to move ahead quickly in selling patent rights and licenses in an attempt to exploit his engine. The General Society's fortunes suffered from the problems with the engine and the failure of the solidarity clauses. By 1906 M.A.N. was virtually running the General Society. Diesel not only lost control over his patents, but he also sacrificed much of his wealth. Suppose he had been content to accept a salaried post as a consulting engineer at M.A.N.? But this was not his choice. His poor health and bad business sense, combined with the problems of the early engine, led to disaster.

The invention, development, and innovation of the diesel engine demonstrate the confluence of a number of factors: the desire for personal fame; the impact of social ideas; the effects of an economic recession; the search, inspired by the advancement of the science of thermodynamics, for a more thermally efficient internal combustion engine; and the luck of obtaining farsighted industrial backing. For these reasons, the Diesel story provides a multicausational explanation of technological change. It reflects the growing interaction of technology with science and industry at the turn of the century. It points up the often difficult transition from developed invention to marketable product as technology became more complex. It in many ways exemplifies the transitional period from the individualistic inventor-entrepreneur based industrialism of the nineteenth century to the more anonymous, mass-based corporate industrialism of the twentieth century. Interwoven among these forces, however, is a personal story of triumph and tragedy. This interplay of individual actors with larger technical, social, and economic forces produces technological change and makes the Diesel story meaningful.

# EPILOGUE

## The Triumph of an Invention

•

The year 1983 marked the 125th anniversary of Rudolf Diesel's birth, which was celebrated, especially in Germany, with newspaper articles and exhibitions. The next year, a new aerospace wing was opened at the Deutsches Museum, in Munich. Mounted near the entrance and serving as a transition from the nearby internal combustion exhibition stands the 1897 diesel engine. Both anniversary and exhibition help recall how far the diesel engine has come since those precarious early days at the beginning of this century.[1]

Already during the last years of Diesel's life, slow but steady improvements were being made in the design of the diesel engine. The replacement of the single-stage air compressor and the sieve vaporizer with the two-stage compressor and the plate vaporizer overcame the most immediate problems and opened the way to a marketable engine. However, many problems continued to exist. For example, the air compressor needed for fuel injection was bulky and lowered the engine's efficiency. Demands of the diesel simply exceeded the technology of the day, and years of experiment and labor were required on the part of numerous engineers working in factories throughout Europe and America before these demands were met. Indeed, progress continues to be made today, especially in the combustion process, including such areas as fuel injection, combustion chamber shape, and air-fuel mixture.[2]

Nonetheless, the diesel engine slowly carved out a niche for itself, especially in areas where fuel efficiency, safe and inexpensive fuel, and dependability played a role.[3] At first, the air compressor and the weight of early engines restricted them to large horsepower applications.[4] They found a place between lower-power gas engines and larger steam turbines. Throughout most of the first decade of the twentieth century, the diesel was primarily used for stationary power generation. Also during this time the marine diesel was being developed. World War I spurred advances in diesels, especially for use in submarines, and after the war an explosion of interest occurred, first in marine applications, and then, because of the introduction of direct fuel injection in the 1920s, in vehicular transportation.

An example of stationary diesel use was in electric-power generation. In 1904 the first diesel electric-power plant in the world opened in Kiev; six M.A.N. engines, whose horsepower totaled 2,400, operated the municipal tram system. In the 1920s one of the peaks of stationary diesel development was a 15,000-horsepower Blohm & Voss engine, which provided electricity for the Neuhof Power Works, in Hamburg. Diesel engines are still used today for electric-power generation, especially in remote areas where an independent power source is needed.[5]

Already in 1901 M.A.N. had laid the basis for lighter, cheaper engines of intermediate horsepower by developing the trunk piston engine. The earliest diesels had been of the crosshead type, copied from steam engines, where a sliding member—the crosshead—served to pin both piston rod and connecting rod. Trunk pistons, taken over from gas engines, were hollow at one end and had the connecting rod pivoted inside the piston. Crosshead engines continued to be used for large stationary and marine engines.[6]

The advantages of diesel engines for marine applications were early recognized. Not only could the boiler and even the smokestack be eliminated, but also fuel consumption could be significantly lowered. As early as 1897 and 1898, marine engines were designed in England and America, though they were never installed in ships.[7] The first diesel-powered boat was the *Petit Pierre*, built by Dyckhoff in France in 1903, which operated along the Rhine-Marne canal. Diesel took part in the test runs and wrote to his wife on October 25, 1903, that he was celebrating the momentous event with an omelet and country wine.

Marine applications followed rapidly on this modest beginning. The problem of reversing engines had been solved by the end of the first decade of the twentieth century. In 1903 and 1904 the Nobel Engine Company, in Russia, constructed engines for two tankers that ran from Astrakhan via the Volga-Marinski canal system and Lake Ladoga to St. Petersburg. The *Vandal* and *Ssmarat* were the first reliable diesel cargo ships.[8]

Even though during the early 1900s neither the French nor the German admiralties showed particular interest in diesel engines, M.A.N. and Krupp together began work on diesel submarine engines in 1901. Besides the economy involved, diesel fuel was much less volatile than gasoline, a major advantage in submarines. Efforts were crowned with success when in 1907 the French Navy installed two M.A.N. 300-hp engines in the submarines *Circe* and *Calypso*. By 1908 the German Navy was ordering test engines, and the outfitting of its U-boats proceeded apace after 1911. Both M.A.N. and Krupp delivered diesel engines for this purpose. Half of all the U-boats during World War I were equipped with M.A.N. four-stroke diesel engines.[9]

A milestone in marine applications was the first oceangoing diesel ship, the 7,000-ton *Seelandia*, built by Burmeister and Wain, in Copenhagen, and powered by two 1,000-hp, eight-cylinder, four-stroke engines. The ship made its maiden voyage to Bangkok in 1912 and caused a storm of interest in shipbuilding circles, including navies.[10]

In December 1911 the Norwegian explorer Roald Amundsen was the first to reach the South Pole. His expedition used the schooner *Fram*, powered by a 180-hp diesel engine that was built by the Swedish A/B Diesels Motorer. Amundsen was later to telegraph the company: "Diesel motor excellent."[11]

Just before and during World War I, both Krupp and M.A.N., in combination with Blohm & Voss, began to construct two-stroke marine engines for both warships and cargo ships. The two-stroke engine combines the intake and compression strokes and the power and exhaust strokes. As the piston approaches bottom dead center, intake and exhaust ports are uncovered. Burnt gases stream out and fresh air enters. After the piston passes bottom dead center and moves up again, it covers the intake and exhaust ports. When the piston reaches top dead center, fuel is injected and a power stroke occurs. The two-stroke engine is more complicated than the four-stroke because it needs to perform operations faster and because proper air-exhaust gas exchange is difficult. Indeed, proper scavenging (sweeping exhaust gases from the cylinder) presented one of the most difficult early problems.

Despite its inherent limitations, the two-stroke engine can generate more power than a four-stroke engine of the same size. Also, it only proved feasible to build double-acting two-stroke engines. Therefore, more horsepower could eventually be attained with large two-stroke engines. In 1967, for example, M.A.N. could build a twelve-cylinder, two-stroke engine that supplied 4,000 hp per cylinder, or a total of 48,000 hp. By contrast, in 1979 the maximum capability of four-stroke engines was eighteen cylinders for 32,400 hp at 1,800 hp per cylinder.[12]

After World War I, diesel marine engines, both four-stroke and two-stroke, were rapidly perfected by a variety of firms. By 1924 some 40 percent of all newly built ship engines were diesel. By 1925 about 70 percent of new ship engines in Germany were diesel.[13] Steam turbines were still often favored for passenger ships because of their smoother ride, but diesels had won the day in cargo vessels.

The real advances in diesel technology during the 1920s, however, were not in increasing engine size but in the technology of direct, or solid, fuel injection. Only when the air compressor used for fuel injection could be dispensed with was the way opened for the development of small high-speed engines that could be adapted to vehicles. As early as 1903, Prosper L'Orange (1876–1939), an engineer at Deutz of French Huguenot background, began to formulate what became known as the precombustion chamber process. In this process, the main combustion chamber is connected by a channel to a smaller chamber. Fuel injected into the channel is driven by the pressurized air into the smaller chamber, where combustion begins. The partially ignited mixture is then swept back into the main combustion chamber, where combustion is completed. Although work on such direct injection systems was suspended during World War I, in the 1920s a number of firms such as Deutz, Benz, where L'Orange then worked, and M.A.N. resumed their effort. By the end of the 1920s, a variety of processes had been developed: direct injection, precombustion chamber, swirl chamber, and air cell.

Other technical improvements complemented the advances in direct injection systems. Between 1922 and 1930, the firm of Robert Bosch, in Stuttgart, which had earlier perfected the electrical ignition system for gas engines, developed a standard fuel-injection pump for diesel engines. As early as 1905, the Swiss engineer Alfred Büchi (1879–1959), working eventually for Sulzer Brothers in Winterthur, began experiments with supercharging, though the process was not fully perfected until the 1920s. Supercharging is a method for precompressing and thus increasing the amount of air admitted to a cylinder for combustion. The result is an increase in horsepower output. The most commonly known form of supercharging today is turbocharging in automobile engines, in which exhaust gases are used to drive a turbine, which in turn is linked to an air compressor that feeds the air-intake manifold.[14]

The technical improvements of the 1920s opened the way for diesel-powered vehicles. During the period of the development of the engine from 1893 to 1897, Diesel had considered its application to locomotives and automobiles. As early as 1897 and 1898, he exchanged ideas about automobile engines with the owner of a Viennese carriage factory, Ludwig Lohner, and with an engineer, Adolf Klose, who worked for the Nuremberg Engine Works. Nuremberg then undertook to construct both locomotive and automobile engines, but these early experiments came to nothing because of technological problems.[15]

Diesel pursued his efforts with both these types of engines between 1904 and 1908, but again these efforts were ahead of their time and failed. In 1912–13 Sulzer Brothers finally produced the first successful diesel locomotive engine in the world. In 1914 five diesels were put into service by the Prussian and Saxon state railroads.[16]

Not until after World War I, however, did diesel-electric locomotives begin to make substantial gains. Diesel locomotives were four times as efficient as steam engines. They could run for long periods without maintenance, accelerate a train more rapidly, run at sustained speeds more easily, and did not need the boiler and water supplies required by steam engines. In the United States, it was primarily General Electric that furthered diesel locomotive building before World War II. By the end of the 1930s, the company was producing diesels at a price cheaper than its competition. Acceptance was at first modest. Between 1925 and 1935, only eighty-seven diesels were in U.S. railroad service. But, between 1936 and 1941, the figure was 1,369. After World War II, diesels completely won the day and by the 1960s railroads in Europe and America were totally electric or diesel-electric, though steam engines could still be found in other parts of the world.[17]

Shortly before his death, Rudolf Diesel is reputed to have written a friend: "I am still of the firm conviction that the automobile engine will come. Then I will consider my life's task completed."[18] Certainly, there is no more obvious symbol of the engine in the popular mind than diesel trucks and automobiles. After long experimentation, in 1923 successful tests of the first diesel truck engines were conducted at Benz and at M.A.N. Both companies were able to exhibit diesel trucks at a Berlin automobile exhibition in 1924. By 1934 some 60 percent of all new trucks in Germany that accommodated more than a two-ton load had diesel engines.[19] In a few years, all truck-building firms inside and outside Germany began to use diesel engines.

Along with improvements in truck engines during the late 1920s and early 1930s, efforts were underway to perfect a viable diesel automobile engine. The Bosch injection pump facilitated these efforts. The first factory to attempt to fit diesel engines into a small, two-seat automobile was the Dorner Ölmotoren A. G., at Hanover, in 1924. Soon experiments were under way in Germany, Austria, France, Switzerland, and the United States.

In the United States it was especially Clessie Cummins, owner of the Cummins Engine Company, in Columbus, Indiana, who pioneered in making Americans "diesel conscious." For example, in 1930, he drove a Packard, outfitted with an adapted diesel ship engine, some 800 miles from Indianapolis to New York City at a fuel cost of only $1.38! In 1935 he drove an Auburn-Cummins diesel from New York City to Los Angeles and incurred fuel expenses of a mere $7.63.

These early attempts all featured adapting a variety of diesel engines to automobiles, but development of diesel engines made especially for this purpose was yet to come. The pioneer in this area, in the early 1930s, was Daimler-Benz in Germany. Its efforts were crowned by success in 1936, when the first true diesel automobile marketed for sale, the Mercedes 260D, was exhibited at the 1936 Berlin automotive exhibition. It was a six-seater whose four cylinders delivered 45 hp. By the end of the 1930s, diesel automobiles were also being made by Hanomag, in Germany, and Peugeot, in France.

After World War II, the growth of the diesel automobile industry began again. Europe led the way, though the 1970s marked a real breakthrough in the United States, especially because of the oil crisis and embargo of the early 1970s and the resulting high gasoline prices. Volkswagen (VW) introduced the "Rabbit" (known in Germany as the "Golf") in 1976; by the end of the 1970s, American manufacturers like Ford and General Motors, and Japanese producers like Nissan and Mitsubishi, had introduced diesel models. Japan's pioneer was Isuzu, part of the Hitachi concern, which had already introduced a diesel automobile in 1961.[20]

The oil glut of the early 1980s and the resulting lower gasoline prices in addition to environmental concerns have temporarily dampened diesel automobile sales, especially in the United States. Although diesel exhaust gas contains less carbon monoxide than gasoline engine exhaust, it does include particulates, such as nitrogen oxide, that are associated with smog creation as well as certain forms of hydrocarbons that some researchers contend may cause cancer. Experiments with recirculating exhaust gases and with more precise ignition timing, producing more complete combustion, have helped to lower emissions, but diesel exhaust still provokes environmental concern.[21] In Germany the environmentalist "Green" Party is actively campaigning for restriction of diesel sales.

The environmental debate of the 1970s called the whole era of fossil fuels into question, and, possibly in the not too distant future, all fossil-fuel-burning internal combustion engines will be a thing of the past. Until then, however, diesel engines are still the most fuel-efficient of all such engines. Advances in supercharging have caused their thermal efficiency to rise to some 42 percent; brake thermal efficiencies are between 32 and 35 percent. Comparative figures for the gas engine average out at 33 percent and 26 percent, respectively.[22] In 1982 a specially designed Peugeot diesel automobile was driven from Detroit to Knoxville at fifty-five miles per hour, averaging seventy-four miles per gallon.[23]

Rudolf Diesel lived only long enough to see the beginnings of the remarkable development and innovation of his engine. His vision outstripped the technical limitations of his day, and in the end he lost both control of his invention and the wealth his engine had once brought him. Perhaps the ultimate irony is that, though "diesel" has come to stand for an engine known the world over, seen on truck stop and service station signs across the United States, the man behind the name is forgotten by most. It is to be hoped that this small contribution will help to right the balance.

# Appendixes

## Appendix 1:
### Chronology of the Invention and Development of the Diesel Engine, 1878–1897

| | |
|---|---|
| 1878–79 | Diesel hears Carl Linde's lectures at the Munich Technische Hochschule on the thermal inefficiency of the steam engine and on the Carnot cycle as the most efficient heat-engine cycle. |
| 1880–89 | Diesel in France as representative of the Linde Refrigeration Company. |
| 1883–89 | Works on the ammonia engine. |
| 1889–early 1892 | Prepares "Theorie und Construction eines rationellen Wärmemotors." |
| 1890 | Moves to Berlin. |
| February 11, 1892 | Writes Linde announcing his new theory. |
| February 27 | Prepares first patent draft. |
| March 7 | Requests developmental support from the Augsburg Engine Works. |
| March 15 | Patent Office rejects application. |
| March 20 | Linde's reply supports Diesel's theory, but declines material support; Linde sends Diesel's manuscript to Professor Moritz Schroeter. |
| March 29 | Schroeter agrees with the theory, but doubts its practicality. |
| April 2 | Augsburg and Heinrich Buz reject Diesel's proposal for support. |
| April 6 | Diesel asks Linde to help build a consortium. |
| April 9 | Diesel writes Buz suggesting lower pressures and playing up economic advantages of the engine. |

| | |
|---|---|
| April 13 | Diesel's letter to Langen requesting support, which is rejected. |
| April 13 | Corresponds with Schroeter suggesting lower pressures; copy sent to Augsburg. |
| April 20 | Augsburg agrees to test engine. |
| May 2 | Schroeter writes Diesel calling attention to closeness of compression and expansion lines in the theoretical indicator diagram. |
| December 23 | Patent No. 67027 granted. |
| January 1893 | *Theorie und Konstruktion* published and sent to various experts. |
| January 19 | Diesel solicits developmental support from Krupp. |
| February 4 | Schroeter's panegyric review of Diesel's book appears in the *Bayrisches Industrie- und Gewerbeblatt*. |
| February 18 | Visit of Krupp directors to Augsburg is probably decisive in winning Krupp support. |
| February 21 | Diesel-Augsburg contract signed. |
| April 10 | Diesel-Krupp contract signed. |
| April 15 | Augsburg-Krupp contract signed. |
| March–April | Otto Köhler criticizes Diesel's ideas in a letter to A. Venator and in a speech to the Cologne VDI (speech published in the *VDI-Zeitschrift* in September). |
| Mid-May to Mid-June | Diesel conceives the idea of constant-pressure combustion; abandons the concept of constant-temperature, isothermal combustion. |
| July–August | First series of tests at Augsburg; compression ignition achieved; engine does not run on its own power. |
| August 1893–January 1894 | Engine is rebuilt. |
| January–April | Second series of tests; air-blast injection of fuel employing a Linde air compressor; kerosene used as fuel. |

| | |
|---|---|
| June–November | Third and fourth testing periods; various ignition devices and fuels tested. |
| November 1894–<br>March 1895 | Construction of second test engine. |
| March–September 1895 | Fifth series of tests; engine runs on its own; kerosene settled on as fuel; tests show mechanical efficiency of 54 percent and brake thermal efficiency of 16.6 percent. |
| October | Krupp may have considered withdrawing from contract. |
| January 1896 | Diesel and Augsburg begin pushing marketability of engine. |
| March–October | Construction of third test engine. |
| October 1896–January 1897 | Sixth and final testing period; sieve atomizer adopted for fuel injection. |
| February | Engine shown to interested parties. |
| February 17 | Schroeter conducts official tests of the engine. |
| June 16 | Diesel and Schroeter introduce the engine to the world at the VDI annual meeting in Kassel. |

## Appendix 2:
### Concluding Remarks from
### Diesel's 1892 Manuscript "Theorie und Construction eines rationellen Wärmemotors"

## Schlussbemerkungen

Gewöhnlich wird nach Zustandekommen einer Erfindung von einiger Bedeutung behauptet, die Sache wäre längst bekannt und es werden allgemeine Äusserungen und Hoffnungen häufig so gedeutet, als enthielten sie die Grundidee der neuen Sache. Es ist deshalb wohl nicht unberechtigt hier mit wenigen Worten den jetzigen Stand der Motorindustrie zu kennzeichnen.

Zunächst muss constatirt werden, dass heute, Anfangs 1892, keine Maschine wirklich existiert, welche auch nur die Hoffnung gestattet, ähnliche therm. Wirkungsgrade zu erzielen, wie sie der vorgeschlagene Motor von vorne herein theoretisch hat. Es wurde ja an anderer Stelle nachgewiesen, dass weder der Gasmotor noch die Feuerluftmaschine eine solche Hoffnung zulassen, da ihr Arbeitsprincip ein falsches ist, und nicht geändert werden kann, wenn man sie nicht ganz aufgiebt.

Weder auf der Pariser Ausstellung 1889 noch auf der Frankfurter Ausstellung 1891 war irgend ein Motor vertreten, welcher von den 3 Typen: Dampfmaschine, Gasmotor, Heissluftmaschine eine Abweichung zeigte.

Trotz dieser feststehenden Thatsache möchte ich auch durch die Aussprüche einiger Autoritäten den heutigen Stand dieser Frage darlegen.

Zeuner gilt in Deutschland als die bedeutendste Autorität auf diesem Gebiete. In seiner Thermodynamik I (1887) sucht er wiederholt den Arbeitswerth eines Brennstoffes zu bestimmen und kommt auf folgende Schlüsse:

S. 388.- "Man stösst aber hier leider auf eine Frage, zu deren Beantwortung der heutige Stand der Thermodynamik noch nicht die Mittel bietet."

S. 389.- "Es ist bei dem heutigen Stande der Thermodynamik wenig Aussicht vorhanden, dass alsbald eine befriedigende Lösung der Frage gelingen sollte, wie die Verbrennung geleitet, d. h. welcher Verlauf der Verbrennungscurve vorliegen müsste, um auf das Maximum der Arbeit zu gelangen."

S. 454.- "Der Versuch einer weiteren Verfolgung des Problems in der angezeigten Richtung (Verbrennung) muss aber wohl zur Zeit noch als verfrüht bezeichnet werden."

In Band II, der 1890 erschien, sind die Schlüsse des Bandes I einfach wieder angeführt und besprochen, ohne jeden Hinweis auf einen Fortschritt. Da Zeuner über die ganze Litteratur dieses Gegenstandes eine seltene Übersicht besitzt, so sind obige Aussprüche wohl als die herrschende Gesammtmeinung anzusehen.

Der vorgeschlagene neue Motor ist auch unabhängig davon, ob die spec. Wärme der Gase variabel ist und von anderen Fragen, von den Theoretikern als Vorbedingungen zur Untersuchung der Verbrennung aufgestellt wurden. Im Grunde ist bei meinen Untersuchungen der Carnot'sche Satz als Resultat anzusehen und dieser ist ja von den Constanten, ja von dem angewendeten Körper überhaupt, gänzlich unabhängig. Zeuner weist übrigens wiederholt (I. S. 341, 342, 363,) darauf hin, dass aus der Anwendung eines Regenerators die Wirkung der Heissluftmaschinen wesentl. verbessert werden kann. Es wurde an anderen Orten gezeigt, dass hierin ein Irrthum liegt.

Als besten Kenner der Gasmotoren in Frankreich gilt Gustave Richard, dessen Urteil als ein sehr umfassendes betrachtet werden darf. Im folgenden sind einige Auszüge aus der Broschüre "Deux communications sur les moteurs à gaz" gegeben; diese Brochüre erschien im November 1891, und kann also wohl als die allerneuste sachgemässe Kundgebung über diese Angelegenheit betrachtet werden.

[Diesel then quotes several sentences in French from Richard's book and continues:]

Diese Urtheile gehen also, kurz gefasst, dahin: Die Verbrennungserscheinung ist noch ein Räthsel; der Regenerator ist das einzige Mittel an den heutigen Verbrennungsmotoren etwaige Verbesserungen zu erzielen. Die vorausgehenden Untersuchungen und der daraus construirte Motor sind daher gewiss nicht aus bestehenden Ansichten entsprungen.

Noch wenige Worte über den Gang der Entwicklung dieser Erfindung mögen zur Bestätigung des Gesagten gestattet sein.

Meine ursprüngliche Idee war, einen Kleinmotor zu machen, welcher stets marschbereit sei und nur durch ein kurzes Heizen nach je 10–12 st. Betriebe gleichsam wieder "aufgezogen" würde. Ich wählte deshalb flüssiges Ammoniak als motorisches Medium und wurde dadurch zur Untersuchung der Ammoniakdämpfe geführt.- Ich that dies in praktischer und theoretischer Hinsicht, machte sehr eingehende Versuche mit Ammoniakdämpfen, Absorption derselben in verschiedenen Flüssigkeiten u.s.w. und construirte auch einen wirklichen Ammoniakmotor. Theorie und Praxis hatten mich schon auf die Überhitzung der Dämpfe geführt; meine Erfahrungen mit dem gebauten kleinen Motor thaten mir in überraschender Weise die Vortheile der Überhitzung kund. Ich stellte nun eine vollständige Theorie einer Dampfmaschine mit hoch überhitzten Amm. Dpfen auf und fand rechnerisch nicht unbedeutende Vortheile vor den heutigen Dampfmaschinen heraus; dabei zeichneten sich diese Motoren durch ausserordentliche Kleinheit, im Vergleich zu unsern jetzigen Dampfmaschinen, aus; was daran lag, dass zur vortheilhaftesten Durchführung des Processes nicht nur hohe Temperaturen sondern auch sehr hohe Drucke angewendet werden mussten. Um nicht irre zu gehen, berechnete ich auch Maschinen, welche hoch überhitzte Wasserdämpfe anwenden würden. Auch hier zeigte sich die Nothwendigkeit, hohe Drucke zu gebrauchen, da nur durch grosse Druckdifferenz beim Expandiren ein grosses Temperaturgefälle effectiv ausnutzbar wird. Es zeigte sich, dass das ganze wissenschaftliche Material, das wir über Dämpfe besitzen, nicht zur Verfolgung des Problems ausreichte. Ich berechnete die Regnault'schen Dampftabellen hypothetisch bis auf sehr hohe Temperaturen weiter und es stellte sich heraus, dass für unsere Verhältnisse der critische Punkt überschritten wurde, so dass man Flüssigkeits-und Gaszustand nicht mehr unterscheiden konnte; dies brachte mich auf die Idee, die Dämpfe als Gase zu betrachten, nur um ihnen rechnerisch näher zu kommen. Dabei entdeckte ich, dass praktisch kein Unterschied zwischen Dampf und Gas bestehe, dass ich also auch Gas, bzw. Luft verwenden könne; ich behielt aber dabei die von den vorhergehenden Untersuchungen stammenden hohen Drucke und hohe Temperaturen bei. Bei hohen Temperaturen war es aber nicht möglich, eine gewöhnliche Verbrennung vortheilhaft auszunutzten. Deshalb kam ich auf den Gedanken, die Verbrennung in der hoch gespannten Luft selbst vorzunehmen. Die Verfolgung dieser Idee führte auf die im vorstehenden mitgetheilte Verbrennungstheorie und auf den Vorschlag des Motors. Ich hatte noch niemals in meinem Leben eine Theorie der Feuerluft- und Gasmotoren studiert; ich that dies erst nachträglich zur Controle meiner Untersuchungen, fand aber das negative Resultat, welches in obigen Aussprüchen Zeuner's und Richard's seinen Ausdruck findet.

Aus diesen letzten Mittheilungen geht auch hervor, warum das endgiltige Resultat mehr als ein Jahrzehnt auf sich warten liess. Es ging demselben voraus ein vollständig neues Studium der überhitzten Dpfe von Wasser und Ammoniak, unter Ausrechnung vieler Hunderte von Zahlenbeispielen und unter Aufstellung einer grossen Zahl neuer Tabellen; ferner eine Serie von Versuchen über Absorption von Ammoniak, welche die Anlage eines ganzen Laboratoriums und einen Aufwand von ca 3 Jahren erforderte, dann eine Theorie zu diesen Versuchen, dann die Ausführung eines wirklichen Ammoniakmotors unter Anwendung verschiedener Arten von Steuerungen, dann eine Untersuchung gesättigter Dämpfe bei Drucken von mehreren 100 Atm. und sehr hohen Temperaturen; dann eine Aufstellung der Verbrennungstheorie, endlich der zeichnerische Entwurf des vorgeschlagenen Motors.

Dieser ganzen, mehr in praktischem Sinne gerichteten Procedur war eine jahrelange, rein theoretische, Untersuchung vorausgegangen, deren Zweck war, die Eigenschaften von Gasen, Dämpfen und Flüssigkeiten durch einheitliche Gesetze auszudrücken, da ich auf diesem Wege allein hoffte, mich dem Ziele zu nähern, die Wärme unserer Brennstoffe rationeller auszunützen.

Berlin, Anfang 1892.

Diesel

# Notes

### Preface

1. Donald E. Thomas, Jr., "Diesel, Father and Son: Social Philosophies of Technology," *Technology and Culture*, 19 (July 1978), 376–393.

### Introduction

1. See the Bibliographical Essay for a discussion of materials used in research for this book and an evaluation of secondary works on Diesel.
2. Hugh G. J. Aitken, *Syntony and Spark: The Origins of Radio* (New York: John Wiley and Sons, 1976), pp. 328–336; and Edward W. Constant II, *The Origins of the Turbojet Revolution* (Baltimore: Johns Hopkins University Press, 1980), pp. 32, 245–246.
3. Thomas P. Hughes, "The Development Phase of Technological Change," and Lynwood Bryant, "The Development of the Diesel Engine," in *Technology and Culture*, 17 (July 1976), 423–431 and 432–446; and F. M. Scherer, "Invention and Innovation in the Watt-Boulton Steam-Engine Venture," *Technology and Culture*, 6 (Spring 1965), 165–187. Referring to the terminology for subdividing the process of technological change, Edward Constant says, "Each author has his own vocabulary for describing these functional stages; most differences seem trivial." See his *Origins of the Turbojet Revolution*, pp. 27 and 278, note 59.
4. Bryant, "Development of the Diesel Engine," p. 446.
5. Aitken, *Syntony and Spark*.
6. Ibid., p. 333.

### Chapter 1

1. A full biography of Diesel has been written by his son Eugen, *Diesel: der Mensch, das Werk, das Schicksal* (Hamburg: Hanseatische

Verlagsanstalt, 1937). See the Bibliographical Essay for an analysis of this book. Much of this chapter admittedly depends on it for biographical details, though it is supplemented by other sources and by the author's own interpretations. A reexamination of many of Eugen's sources shows that he essentially quoted them correctly, though sometimes selectively, and that his interpretation colored his reading of them. Interestingly enough, though Eugen, while writing the biography, discussed it with many people, or at least solicited information from them, he never discussed it with his sister, Hedy. Hedy von Schmidt, "Lebenserinnerungen," p. 27, unpublished manuscript in the possession of Hedy's daughter, Frau Dorette Breig, Stockholm-Sollentuna, Sweden.

2. Eugen Diesel, *Diesel*, pp. 170–171.

3. Hedy von Schmidt, "Lebenserinnerungen," p. 13.

4. Eugen Diesel, *Diesel*, p. 37.

5. Lynwood Bryant has emphasized these points to the author. The fact that Diesel wrote in the Latin script is a boon for historians because the old Germanic script is quite difficult for Americans to decipher and even young Germans have trouble reading it.

6. Eugen Diesel, *Diesel*, p. 17.

7. Ibid., pp. 21–28.

8. The M.A.N. Werkarchiv contains a typed set of answers by Rudolf's younger sister, Emma Barnickel, to questions Eugen Diesel asked her in 1935–36. Some of this material was the basis for Eugen's description of his father's early days in Paris. Although Emma does mention hard times, she does not dwell on this aspect to the extent Eugen does. She says the family was willing to make sacrifices, especially for the eldest daughter, Louise, who was musically gifted.

9. Rudolf Diesel's letters from this period indicate that his sister's sudden death profoundly shocked him and his family. Interestingly enough, however, none of the evidence indicates the cause of death. Eugen Diesel's biography also does not answer this question.

10. A copy is in the Northwestern University Library, Evanston, Illinois.

11. Eugen Diesel, *Diesel*, p. 128.

12. Ibid., p. 34.

13. Ibid., pp. 32–33.

14. Ibid., pp. 34–35.

15. Ibid., pp. 39–40.

16. Professor Barnickel later married Diesel's younger sister Emma, after the death of his first wife.

17. Eugen Diesel, *Diesel*, p. 73.

18. Ibid., pp. 74–75, 84.

19. This patent draft, entitled "Neue rationelle Wärmekraftmaschine" and dated February 26, 1892, is in the M.A.N. Werkarchiv.

20. Eugen Diesel, *Diesel*, pp. 86–88.

21. Ibid., pp. 91–92.

22. Ibid., p. 89; and letter of November 22, 1874, in the M.A.N. Werkarchiv.

23. Theodor Diesel to Rudolf, January 31, 1875, M.A.N. Werkarchiv.

24. Theodor Diesel to Professor Barnickel, March 31, 1875, ibid.

25. A copy of Diesel's *Inscriptionsbuch* is in the M.A.N. Werkarchiv.

26. See Carl Linde, *Aus meinem Leben und von meiner Arbeit* (Munich: R. Oldenbourg, 1916), p. 32.

27. Eugen Diesel, *Diesel*, pp. 93–94, 98–99.

28. Linde, *Aus meinem Leben*, p. 53.

29. Eugen Diesel, *Diesel*, pp. 118–123.

30. From a copy of Diesel's sister Emma's journal, in the M.A.N. Werkarchiv. During this period, the Technische Hochschulen could not confer degrees equal to those of the older universities. See chapter 2.

31. Eugen Diesel, *Diesel*, pp. 124–125.

32. Letter in the M.A.N. Werkarchiv. The letter is written in English. Eugen Diesel quotes this letter with a slight variation in his *Diesel*, p. 143.

33. Diesel to Martha, June 24, 1883, M.A.N. Werkarchiv.

34. Eugen Diesel, *Jahrhundertwende, gesehen im Schicksal meines Vaters* (Stuttgart: Reclam, 1949), pp. 50–55.

35. Letter to Martha, October 18, 1883, M.A.N. Werkarchiv.

36. Diesel to parents, undated, ibid.

37. Eugen Diesel, *Diesel*, p. 152.

38. Diesel to Carl Linde, February 11, 1892, Diesel *Nachlass*, Deutsches Museum.

39. See chapter 3 for a discussion of the ammonia vapor engine.

40. His sketch, "Idee einer von der Sonne bewegten Maschine," is in the M.A.N. Werkarchiv. Eugen Diesel summarizes it on p. 160 of his *Diesel*.

41. Letters in M.A.N. Werkarchiv. See also Eugen Diesel, *Diesel*, p. 165.

42. Letters in M.A.N. Werkarchiv.

43. Eugen Diesel, *Diesel*, p. 354.

44. The author has discussed Diesel's personality with a physician and psychiatric social worker. They both agreed that his symptoms strongly suggest a manic-depressive condition. While he was writing the biography of his father, Eugen Diesel submitted his father's handwriting to a graphologist. One of the findings was that Diesel's personality was basically manic-depressive. Lena Mayer-Benz, "Deutung der Schrift des Erfinders Diesel," February 26, 1934, in the M.A.N. Werkarchiv. Hedy von Schmidt noted her father's psychology in her memoirs. She said his routine was incredibly regulated. He lived by the clock, allowing himself only a short midday pause and an afternoon walk. Even on vacations he took along his secretary; his mind was always working. Despite these characteristics, Hedy, perhaps understandably, refused to believe that her father was mentally ill (*geisteskrank*). She maintained that he was worn out from years of work and struggle with his enemies. She also claimed that Eugen did not believe his father was mentally ill, but that his biography suggested such an idea—as a number of people told her. When she reported this to Eugen, he professed to be amazed that he could have been so misunderstood. "Lebenserinnerungen," p. 27. Later in life, when he himself was ill, Eugen recognized a streak of instability in his father—and himself. Comments to the author by Lynwood Bryant, who talked with Eugen in 1967.

45. Letter in M.A.N. Werkarchiv.

46. Letter to Martha, July 15, 1888, ibid.

47. Diesel to his father, February 24, 1889, ibid.

48. Letter of April 1, 1889, ibid. Eugen Diesel even saw a similarity between the curve of the Eiffel Tower and the theoretical indicator diagram of the diesel engine! See his *Jahrhundertwende*, pp. 63, 73.

49. Rudolf Diesel, "Abhandlung über den Ammoniak-motor, System Diesel, 1880–1887/88," p. 130, unpublished handwritten manuscript in the M.A.N. Werkarchiv.

50. Eugen Diesel, *Diesel*, pp. 132–133.

51. Ibid., p. 129.

52. Letters to Martha of November 7 and 9, 1889, and February 26, 1890, M.A.N. Werkarchiv.

53. Berlin: Julius Springer, 1893.

54. Both Diesel's and Schroeter's talks are reprinted in Rudolf Diesel, "Diesels rationeller Wärmemotor," Sonderabdruck aus der *Zeitschrift des Vereins Deutscher Ingenieure*, 1897 (Berlin 1897); quote is on p. 19.

55. Letter to Martha, November 2, 1895, M.A.N. Werkarchiv.

56. Eugen Diesel, *Diesel*, pp. 352, 354. Diesel was stricken by an attack of gout during the move into the villa. Such attacks became periodic in later years.

57. The Diesel villa is described in Eugen Diesel's works *Diesel*, pp. 350–354, and *Jahrhundertwende*, pp. 185–201.

58. The German term *Kinderkrankheiten* was often used at this time to describe technical problems with the engine.

59. In 1898 Heinrich Buz's Augsburg Engine Works merged with the Nuremberg Engine Works to form what is today known as the Augsburg-Nuremberg Engine Works (Maschinenbau Augsburg-Nürnberg, or M.A.N.).

60. Eugen Diesel, *Diesel*, pp. 362–363, 403–404. By the time of his death, Diesel even owned twelve homes in Hamburg. "Der Nachlass Dr. Diesels," *Münchener Zeitung*, October 15, 1913, M.A.N. Werkarchiv. Eugen Diesel claims that a clever speculator was able to contact his father during his sanitarium stay in 1898–99 and convince him to put money into construction sites. This was supposedly the beginning of Diesel's financial ruin. Eugen Diesel, *Diesel*, pp. 318–319, *Jahrhundertwende*, pp. 158, 285.

61. Eugen Diesel, *Jahrhundertwende*, p. 192.

62. Ibid., pp. 417–421. The exact story of Diesel's finances is somewhat obscure and probably cannot be reconstructed today. Much of the evidence may have been destroyed when he burned many of his private papers in September 1913.

63. N.A.G. stands for the Neue or Nationale Automobil-Gesellschaft, an automobile company founded in 1901 by the Allgemeine Electrizitäts-Gesellschaft, or German General Electric Company.

64. Eugen Diesel, *Diesel*, p. 394.

65. Dorette Brieg, Diesel's granddaughter, to the author, February 9, 1984.

66. Hedy von Schmidt, "Lebenserinnerungen," pp. 60–81. The reason why Arnold was unhappy working for his father-in-law cannot be ascertained.

67. Eugen to Rudolf, June 4, 1909, and Diesel to Martha, July 13, 1911, in the Eugen Diesel *Nachlass*, Freiburg.

68. For Eugen Diesel's life and writings, see the author's article, "Diesel, Father and Son," pp. 376–393.

69. Munich and Berlin: R. Oldenbourg, 1903.

70. The diary is in the Diesel *Nachlass*, Deutsches Museum.

71. Eugen Diesel, *Diesel*, p. 380.

72. Ibid., pp. 381–382.

73. A good example is *Scientific American*'s reprint of a speech Diesel made in England, just before coming to America. In it, he stressed that the use of oil as a fuel would open up new areas of the world to industrialization. He even mentioned peanut and other vegetable oils as possible fuels of the future. See "The Diesel Oil-Engine and Its Industrial Importance: The Simplest and Most Efficient of Motors," *Scientific American*, April 20, 1912, pp. 357–358.

74. The Diesel *Nachlass*, Deutsches Museum, contains Diesel's diary and all his correspondence relating to the 1912 trip. Included are a number of newspaper articles on the *Titanic* sinking that he clipped while in America.

75. Eugen Diesel, *Diesel*, p. 411. Material in the Edison Papers, in West Orange, New Jersey, does not illuminate this meeting from Edison's point of view.

76. From Diesel's 1912 diary in the Deutsches Museum. This diary is much less extensive than the 1904 diary. One of Diesel's few remarks about American society in the former is: "The American principle of life from top to bottom is 'the highest output for the lowest return.' Everyone is exploited by someone else."

77. Diesel's lecture was later expanded and turned into the book *Die Entstehung des Dieselmotors*, which was published by Springer in Berlin shortly before Diesel's death in September 1913. Riedler, too, would later publish his arguments in *Dieselmotoren: Beiträge zur Kenntnis der Hochdruckmotoren* (Vienna, Berlin, London: Verlag für Fachliteratur, 1914).

78. Diesel corresponded a good deal with Springer in August and September 1913 about his own book, *Die Entstehung des Dieselmotors*, which was about to appear. Springer informed Diesel of Lüders's impending publication, and Diesel wrote Springer a number of letters requesting information on Lüders's background, the nature of the book, and time of

its publication in relationship to his own work. Springer advised Diesel not to reply to Lüders's attack because, in an argument, Lüders was a malicious and obstinate opponent who always had the last word. The author wishes to thank Dr. Michael Davidis, of the Deutsches Museum Research Institute, who made the Diesel-Springer correspondence available.

79. Diesel, *Entstehung*, pp. 151–152.

80. Eugen Diesel, *Diesel*, pp. 429–430.

81. Ibid., pp. 431–432.

82. Eugen Diesel, *Jahrhundertwende*, p. 290.

83. Eugen Diesel, *Diesel*, pp. 442–443.

84. Ibid., p. 448.

85. Letter of September 28, 1913, M.A.N. Werkarchiv.

86. Eugen Diesel, *Jahrhundertwende*, pp. 297–299.

87. Eugen Diesel raises but then rejects the idea that his father deliberately committed suicide to take attention away from his difficulties, disarm his critics, and ensure the success of his name and engine. See his *Jahrhundertwende*, pp. 289–290. Immanuel Lauster was one of the M.A.N. engineers who was most responsible for the success of the diesel engine, though he and Diesel did not get along. He maintained that Diesel may have killed himself to create a martyr's image and take attention away from the fact that it was not he, but M.A.N., who created the working diesel engine. Perhaps Lauster's comments were made primarily because of his dislike for Diesel. Immanuel Lauster, "Der Dieselmotor," p. 457, unpublished manuscript in the M.A.N. Werkarchiv. On Lauster, see chapters 4 and 5.

88. Eugen Diesel, *Diesel*, p. 464.

89. Martha Diesel lived another thirty years, never able to understand why her husband had not confided in her about his troubles. The Munich villa and almost all its contents were auctioned off to help pay his debts. According to Hedy von Schmidt, friends of the family were asked to purchase some items and give them back, but no one did so. Some things were even stolen from the house. Eventually, Martha received a stipend from German industry that allowed her to live quite adequately. She died in April 1944. Hedy von Schmidt, "Lebenserinnerungen," p. 90.

90. See Joseph Rossman, *Industrial Creativity: The Psychology of the Inventor* (New ed., New Hyde Park, N.Y.: University Books, 1964), pp. 35–55.

91. Ibid., pp. 200–201.

92. Elting G. Morison, *Men, Machines, and Modern Times* (Cambridge, Mass.: MIT Press, 1966), p. 9.

93. Letter of February 20, 1898, in M.A.N. Werkarchiv.

94. Diesel to Eugen, September 25, 1910, Eugen Diesel *Nachlass*, Freiburg.

## Chapter 2

1. American engineers, too, were concerned with status in their society. They believed technological advance led to social progress and that engineers were a mediating force between capital and labor. Many proposed technocratic solutions, but the majority rejected radical social reform. See Edwin Layton, Jr., *The Revolt of the Engineers: Social Responsibility and the American Engineering Profession* (Cleveland and London: Case Western Reserve University Press, 1971).

2. Ludolf von Mackensen, "Der Übergang Deutschlands zu einem technisch führenden Land, 1850–1914," unpublished paper in the author's possession. Numerous general surveys of the German industrial revolution have been written. See, for example, David S. Landes, *The Unbound Prometheus: Technological Change and Industrial Development in Western Europe from 1750 to the Present* (London: Cambridge University Press, 1969); and Hermann Kellenbenz, *Deutsche Wirtschaftsgeschichte*, Vol. II: *Vom Ausgang des 18. Jahrhunderts bis zum Ende des zweiten Weltkriegs* (Munich: C. H. Beck, 1981).

3. The history of the Technische Hochschulen is briefly reviewed in Karl-Heinz Manegold, "Das Verhältnis von Naturwissenschaft und Technik im 19. Jahrhunderts im Spiegel der Wissenschaftsorganisation," in Wilhelm Treue and Kurt Mauel (eds.), *Naturwissenschaft, Technik, und Wirtschaft im 19. Jahrhundert*, Part I (Göttingen: Vandenhoeck und Ruprecht, 1976), pp. 253–283. See also Manegold's *Universität, Technische Hochschule, und Industrie: Ein Beitrag zur Emanzipation der Technik im 19. Jahrhundert* (Berlin: Duncker und Humblot, 1970), and his essay "Technology Academised: Education and Training of the Engineer in the Nineteenth Century," in Wolfgang Krohn, Edwin Layton, Jr., and Peter Weingart (eds.), *Sociology of the Sciences*, Vol. II: *The Dy-*

*namics of Science and Technology* (Dordrecht, Holland: D. Reidel, 1978), pp. 137–158.

4. Charles Coulston Gillispie et al. (eds.), *Dictionary of Scientific Biography* (New York: Charles Scribner's Sons, 1970–80).

5. Mackensen, "Übergang Deutschlands," pp. 6–7.

6. Ludwig Brinkmann, *Der Ingenieur* (Frankfurt/M: Literarische Anstalt Rütten and Loening, 1908), pp. 9–18. Brinkmann's work was part of a series of sociopsychological monographs that were edited by Martin Buber and entitled *Die Gesellschaft*.

7. A major sociological study of the VDI and its relationship to social questions is Gerd Hortleder, *Das Gesellschaftsbild des Ingenieurs: Zum politischen Verhalten der technischen Intelligenz in Deutschland* (Frankfurt/M: Suhrkamp, 1970). See also the newer study: Karl-Heinz Ludwig and Wolfgang König (eds.), *Technik, Ingenieure, und Gesellschaft: Geschichte des Vereins Deutscher Ingenieure, 1856–1981* (Düsseldorf: VDI-Verlag, 1981).

8. Fritz Stern, "The Political Consequences of the Unpolitical German," in *The Failure of Illiberalism* (New York: Alfred A. Knopf, 1972), pp. 3–25. The relationship of the newer technical colleges to the older university system is discussed passim in Fritz K. Ringer, *Education and Society in Modern Europe* (Bloomington: Indiana University Press, 1979); and Charles E. McClelland, *State, Society, and University in Germany, 1700–1914* (Cambridge: Cambridge University Press, 1980). See also James C. Albisetti, *Secondary School Reform in Imperial Germany* (Princeton: Princeton University Press, 1983). In general, these works support the idea of a split between the older, humanistic universities and the rising Technische Hochschulen.

9. See Heinrich Popitz et al., *Technik und Industriearbeit* (Tübingen: J. C. B. Mohr, 1957), pp. 1–26, for a good discussion of such antitechnological ideas.

10. Manegold, "Technology Academised," pp. 156–157.

11. Cornelis W. R. Grispen, "Selbstverständnis und Professionalisierung deutscher Ingenieure: Eine Analyse der Nachrufe," *Technikgeschichte*, 50, Nr. 1 (1983), especially 42–50. He bases his argument on a statistical analysis of 651 obituaries of engineers appearing in the *VDI-Zeitschrift* between 1870 and 1940. He admits his sample may be biased toward the elite of the profession and warns that his results must be used with caution.

12. The following synopsis of von Weber's ideas is drawn from "Die Stellung der Techniker im staatlichen und socialen Leben," *Wochenschrift des österreichischen Ingenieur-und Architekten-Vereins*, 2, No. 7 (February 17, 1877), 59–60; and No. 9 (March 3, 1877), 85–88.

13. Grispen, "Selbstverständnis und Professionalisierung," p. 42. For the following argument, see Hortleder, *Gesellschaftsbild*, pp. 18–37.

14. Hortleder, *Gesellschaftsbild*, p. 37. Grispen, "Selbstverständnis und Professionalisierung," p. 58, argues that professors may also have become role models during the last half of the nineteenth century, based on the large number of their obituaries appearing in the *VDI-Zeitschrift*. He also indicates the role model may have occurred because of the academicians' leading role in the engineering profession. Presumably, he is referring to their lead in the fight for social status.

15. Brinckmann, *Der Ingenieur*, pp. 73–74.

16. Hortleder, *Gesellschaftsbild*, pp. 37–49.

17. Ibid., pp. 44–46. Hortleder even regards technocracy as a type of neutralization solution, where engineers or the use of engineering principles in running the state would neutralize social conflicts. Ibid., p. 103. See also Karl-Heinz Ludwig, *Technik und Ingenieure im dritten Reich* (Düsseldorf: Droste Verlag, 1974), pp. 20–27. Hortleder's thesis must be looked upon as an intriguing first attempt in what Manegold calls "a largely untilled field in social history . . . in many ways we are more knowledgeable and better informed about the peasant of the fifteenth and sixteenth century than about the engineer of the nineteenth century." See his "Technology Academised," p. 152. In a recent essay, Wolfgang König criticizes what he views as Hortleder's leftist ideological approach and questions some of his methodology, especially in determining the social composition of the VDI and its leadership. See König's "Die Ingenieure und der VDI als Grossverein in der Wilhelminischen Gesellschaft, 1900–1918," in Ludwig and König (eds.), *Technik, Ingenieure, und Gesellschaft*, pp. 261, 284–285, note 123.

18. Brinkmann, *Der Ingenieur*, pp. 73–83.

19. Ibid., p. 89.

20. Wilhelm Treue, "Ingenieur und Erfinder: Zwei sozial- und technikgeschichtliche Probleme," *Vierteljahrsschrift für Sozial- und Wirtschaftsgeschichte*, 54 (1967), especially 472.

21. See Karl Dietrich Born, "Der soziale und wirtschaftliche Strukturwandel Deutschlands am Ende des 19. Jahrhunderts," in Hans-Ulrich

Wehler (ed.), *Moderne Deutsche Sozialgeschichte* (Cologne-Berlin: Kiepenheuer and Witsch, 1966), pp. 171–184. On the thesis of the feudalization of the German middle class, see, for example, Hans Rosenberg, *Grosse Depression und Bismarckzeit: Wirtschaftsablauf, Gesellschaft, und Politik in Mitteleuropa* (Berlin: Walter de Gruyer, 1967), p. 150; and Gordon A. Craig, *Germany, 1866–1945* (New York: Oxford University Press, 1978), pp. 99–100. The thesis that Wilhelmine Germany was dominated by antimodern elites, of which the feudalization of the middle class is one by-product, has recently been challenged by a group of younger, primarily British, historians. See, for example, Richard J. Evans, "Wilhelm II's Germany and the Historians," pp. 11–36, and David Blackbourn, "The Problem of Democratisation: German Catholics and the Role of the Centre Party," especially pp. 178–179, both essays in Richard J. Evans (ed.) *Society and Politics in Wilhelmine Germany* (London: Croom Helm, 1978). See also Theodore S. Hamerow, "Guilt, Redemption, and Writing German History," *American Historical Review*, 88 (February 1983), 68–71.

22. Quoted in König, "Die Ingenieure und der VDI als Grossverein," in Ludwig and König (eds.), *Technik, Ingenieure, und Gesellschaft*, p. 252.

23. See Hans-Joachim Braun, "Ingenieure und soziale Frage, 1870–1920," *Technische Mitteilungen*, 73 (October 1980), 793–798; and (November–December 1980), 867–874.

24. Eugene C. McCreary, "Social Welfare and Business: The Krupp Welfare Program, 1860–1914," *Business History Review*, 42 (Spring 1968), 24–45; and Peter Batty, *The House of Krupp* (New York: Stein and Day, 1966), pp. 86–92.

25. His attempts were only partially successful by 1906. See Erich von Kurzel-Runtscheiner, "Heinrich Ritter von Buz: Ein Führer der süddeutschen Industrie," pp. 93–103, unpublished manuscript in the M.A.N. Werkarchiv. The manuscript dates from World War II. A version published after the author's death in *Lebensbilder aus den bayrischen Schwaben*, 10 (1973), 319–360, omits the section on Buz's social ideas.

26. Walther Peter Fuchs, "Die geschichtliche Gestalt Ferdinand Redtenbachers," *Zeitschrift für die Geschichte des Oberrheins*, 107 (1959), 205–222. Apparently, Buz, one of Redtenbacher's students, did not absorb his teacher's cultural theories. On Redtenbacher and the Technische Hochschulen, see also the important work of Gustav Goldbeck, *Technik*

*als geistige Bewegung in den Anfängen des deutschen Industriestaats* (first published 1934; Düsseldorf: VDI Verlag, 1968), pp. 13, 31-38, 74-75. Goldbeck points out that the idea of technological progress stimulating social advancement comes from the eighteenth-century Enlightenment. Nineteenth-century liberals often regarded technology as a progressive force and advocated its advancement.

27. Alois Riedler, *Unsere Hochschulen und die Anforderungen des zwanzigsten Jahrhunderts* (Berlin: A. Seydel, 1898), pp. 50-70.

28. Riedler's social ideas are analyzed in Volker Hunecke, "Der 'Kampf ums Dasein' und die Reform der technischen Erziehung im Denken Alois Riedlers," in Reinhard Rürup (ed.), *Wissenschaft und Gesellschaft: Beiträge zur Geschichte der technischen Universität Berlin, 1879-1979* (Berlin: Springer Verlag, 1979), I, 301-313.

29. On Siemens, see his *Inventor and Entrepreneur: Recollections of Werner von Siemens* (London: Lund and Humphries, 1966), especially pp. 266-268; Georg Siemens, *Der Weg der Elektrotechnik: Geschichte des Hauses Siemens* (Freiburg and Munich: Karl Alier, 1961), I, 78-79, and II, 287-288; and Jürgen Kocka, *Unternehmensverwaltung und Angestelltenschaft am Beispiel Siemens, 1847-1914* (Stuttgart: Ernst Klett, 1969). On Harkort, see Goldbeck, *Technik als geistige Bewegung*, pp. 80, 104, 121-122; on Lilienthal, Gerhard Halle, *Otto Lilienthal* (2nd ed.; Düsseldorf: VDI-Verlag, 1956); on Abbe, Ingrid-Bauert Keetman, *Deutsche Industriepioniere* (Tübingen: Rainer Wunderlich, 1966), pp. 149-151; On Ochelhäuser, Wolfgang von Geldern, "Wilhelm Ochelhäuser als Unternehmer, Wirtschaftspolitiker, Sozialpolitiker, und Kulturpolitiker," Dissertation: Technische Universität, Hanover, 1970, pp. 132-149. The dissertation by David Lee Cahan, "The Physikalisch-Technische Reichsanstalt: A Study in the Relations of Science, Technology, and Industry in Imperial Germany," Johns Hopkins, 1980, discusses Siemens's relations to the founding of the PTR; see especially pp. 78-99.

30. See Schmoller's *Zur Geschichte der deutschen Kleingewerbe im 19. Jahrhundert* (Halle: Verlag der Buchhandlung des Waisenhauses, 1870), p. 702. On the discussion in engineering circles of the small power source for industry, see Braun, "Ingenieure und Soziale Frage," pp. 796-798.

31. Wilhelm Treue, *Eugen Langen und Nic. August Otto: Zum Verhaltnis von Unternehmer und Erfinder, Ingenieur und Kaufmann* (Munich: F. Bruckmann, 1963), pp. 16, 54-55.

32. His decentralization ideas are discussed in his *Lehrbuch der Kinematik*, Part I: *Grundzüge einer Theorie des Maschinenwesens* (Braunschweig: Friedrich Vieweg und Sohn, 1875), pp. 514–530. See also Eugene Ferguson's introduction to the English translation *Kinematics of Machinery: Outline of a Theory of Machines*, trans. and ed. Alexander B. W. Kennedy (New York: Dover, 1963), p. vii, from which the quotes are taken. The biographical sketch is drawn from pp. v–x. Reuleaux's theories have been analyzed by Popitz, *Technik und Industriearbeit*, pp. 14–15; Braun, "Ingenieure und soziale Frage," pp. 867–868; and Ernst Kapp, *Grundlinien einer Philosophie der Technik* (Braunschweig: George Westermann, 1877), pp. 198–201.

33. Like Brinkmann, Reuleaux shares the common assumption at the time that the industrial revolution and its social organization fundamentally depended on steam power. Steam "provides a unifying metaphor for modernization." Recent research on the early industrial revolution in England and France demonstrates that traditional sources of power, such as animal and water power, were more important for early industrialization than was once thought. See Dolores Greenburg, "Reassessing the Power Patterns of the Industrial Revolution: An Anglo-American Comparison," *American Historical Review*, 87 (December 1982), 1237–1261; quote is on p. 1239.

34. Braun, "Ingenieure und soziale Frage," p. 867, points out the similarity of Reuleaux's argument to Marx's, as well as the disparity of their solutions.

35. Reuleaux, *Lehrbuch*, p. 523.

36. Ibid., p. 514.

37. Ibid., p. 526.

38. Ibid., p. 529.

39. Ibid., p. 530.

40. Braun, "Ingenieure und soziale Frage," p. 868. See also Braun's article, "Ingenieurwissenschaft und Gesellschaftspolitik: Das Wirken von Franz Reuleaux," in Rürup (ed.), *Wissenschaft und Gesellschaft*, I, especially 296–298.

41. Ibid.

42. Popitz, *Technik und Industriearbeit*, p. 15.

43. Diesel to Martha, February 15, 1898, M.A.N. Werkarchiv; and Eugen Diesel, *Diesel*, p. 290.

44. Eugen Diesel, *Diesel*, pp. 104, 122–123.

45. Ibid., pp. 133–134; and Diesel to parents, June 22, 1880, M.A.N. Werkarchiv.

46. Rudolf Diesel, "Leitende Principien bei Anlage einer Fabrik," document in the M.A.N. Werkarchiv.

47. Hedy von Schmidt, "Lebenserinnerungen," pp. 45–46.

48. From Diesel's letter to Martha of May 16, 1895, quoted in the Diesel "Familienkronik," in the possession of Rainer Diesel.

49. Eugen Diesel, *Diesel*, p. 176.

50. Hedy von Schmidt, "Lebenserinnerungen," p. 27.

51. Eugen Diesel, *Jahrhundertwende*, p. 287.

52. Eugen Diesel, *Diesel*, p. 288; and Diesel to Martha, February 13, 1898, M.A.N. Werkarchiv.

53. Eugen Diesel, *Diesel*, pp. 373–374.

54. Diesel to Martha, January 12, 1907, M.A.N. Werkarchiv.

55. The other motive for invention, the desire to invent a more thermally efficient engine than the steam engine, will be discussed in chapter 3.

56. It was Linde who awoke in Diesel the desire to build a more thermally efficient engine. See chapter 3.

57. Rudolf Diesel, "Anwendungen des Kleinmotors," May 1887, document in the M.A.N. Werkarchiv.

58. Diesel, "Neue rationelle Wärmekraftmaschine," ibid.

59. Diesel, *Theorie und Konstruktion*, p. 89.

60. Ibid., pp. 89–91.

61. Diesel, "Diesels rationeller Wärmemotor," p. 11.

62. Albert Johanning, *Der Dieselmotor: seine Entwicklung und volkswirtschaftliche Bedeutung* (Nuremberg: Wilh. Tümmel, 1901. See also chapter 5.

63. "Die 'Petroleum-Petitionen' der Jahre 1898, 1900, and 1904," unpublished manuscript in the M.A.N. Werkarchiv.

64. Gustav Goldbeck, "Geschichte des Verbrennungsmotors," Part III: "Der Dieselmotor 1900 bis 1960," *Automobil-Industrie*, Nr. 1 (1972), p. 37.

65. Gustav Goldbeck, *Gebändigte Kraft: Die Geschichte der Erfindung des Otto-Motors* (Munich: Heinz Moos, 1965), pp. 110–126.

66. Because of the current "small is beautiful" interest as well as the predictions that advances in computers may decentralize work, education, and even shopping, the advocates of decentralization theories may yet come into their own.

67. Eugen Diesel, *Diesel*, pp. 366–367, *Jahrhundertwende*, pp. 61–62.

68. Diesel to Linde, February 11, 1892, Diesel *Nachlass*, Deutsches Museum.

69. Diesel, *Solidarismus*, p. 68.

70. Ibid., pp. 37–69.

71. Ibid., pp. 25–26; and Diesel's "Erläuternde Ergänzungen zu meiner Druckschrift *Solidarismus*," unpaginated manuscript in the M.A.N. Werkarchiv.

72. Diesel, "Erläuternde Ergänzungen." A picture of the title page of Diesel's own copy of *Solidarismus* is in Eugen Diesel, *Jahrhundertwende*, p. 210.

73. Rudolf Diesel, "Die natürliche Religion: erste Grundgedanken zu einem Werke hierüber," pp. 1–2, manuscript in the M.A.N. Werkarchiv.

74. Ibid., pp. 3–4; and Diesel, "Erläuternde Ergänzungen."

75. Diesel, "Die natüraliche Religion," pp. 3–5.

76. Frank E. Manuel, *The Prophets of Paris* (New York: Harper Torch Books, 1965), pp. 116, 143.

77. Eugen Diesel *Nachlass*, Freiburg.

78. Eugen Diesel, *Jahrhundertwende*, pp. 50–55.

79. In Braun's article "Ingenieure und soziale Frage," p. 869, he mentions the profusion of Christian terminology in *Solidarismus*, but does not place it in the wider scope of Diesel's religious ideas.

80. Diesel, "Erläuternde Ergänzungen."

81. Eugen Diesel, *Diesel*, pp. 384–386, *Jahrhundertwende*, pp. 209–211.

82. Milič Čapek, "Ostwald, Wilhelm," in Paul Edwards et al. (eds.), *The Encyclopedia of Philosophy* (New York: The Macmillan Co. and The Free Press, 1967), VI, 5–6.

83. Diesel, "Erläuternde Ergänzungen."

84. Eugen Diesel, *Jahrhundertwende*, pp. 253–255; and Ostwald to Diesel, May 14, 1913, M.A.N. Werkarchiv.

85. Unfortunately, all that is left of these letters are Diesel's handwritten notes in the M.A.N. Werkarchiv. See Kropotkin's *Mutual Aid: A Factor in Evolution* (New York: New York University Press, 1972); and Martin A. Miller, *Kropotkin* (Chicago and London: University of Chicago Press, 1976), pp. 172–173, 196–197. The Tolstoy comments are in Diesel's "Erläuternde Ergänzungen."

86. On Bellamy, see Sylvia E. Bowman, *The Year 2000: A Critical Biography of Edward Bellamy* (New York: Bookman Associates, 1958); Franz X. Riederer, "The German Acceptance and Reaction," in Sylvia E. Bowman, et al., *Edward Bellamy Abroad: An American Prophet's Influence* (New York: Twayne Publishers, 1962), pp. 151–205; and John L. Thomas, *Alternate America: Henry George, Edward Bellamy, Henry Demarest Lloyd and the Adversary Tradition* (Cambridge, Mass., and London: Belknap/Harvard University Press, 1983).

87. See Diesel, *Solidarismus*, p. 45, for this sentence. Diesel's comments about Bellamy are from his "Erläuternde Ergänzungen." The American edition of Bellamy's *Equality* (New York: D. Appleton and Co., 1897), Chapter XIII, "Private Capital Stolen from the Social Fund," pp. 87–91, discusses the superiority of social over individual production, but does not contain a sentence exactly like the one Diesel claims to have borrowed.

88. G. Briefs, "Pesch, Heinrich," in *Encyclopedia of the Social Sciences*, 1934, XII, 91–92; and Braun, "Ingenieure und Soziale Frage," p. 869.

89. The notes are in the M.A.N. Werkarchiv. Diesel's comments on the French solidarists are from his "Erläuternde Ergänzungen." Eugen Diesel comments that, though his father had read Bourgeois, the similarity in names was accidental. See Eugen Diesel, *Diesel*, p. 367, note. This comment would seem to be inaccurate. Indeed, Eugen's biography, though relating his father's ideas to general currents of the time, rarely discusses specific influences.

90. John A. Scott, *Republican Ideas and the Liberal Tradition in France, 1879–1914* (New York: Columbia University Press, 1951), pp. 124–125, 157–183.

91. Notes on Diesel's *Solidarismus*, in the M.A.N. Werkarchiv.

92. Diesel, *Solidarismus*, pp. 1–2, 50, and "Erläuternde Ergänzungen."

93. Diesel, *Solidarismus*, pp. 2–5; Diesel's notes on a conversation with Brentano on December 8, 1903, in the M.A.N. Werkarchiv; Thomas, *Alternate America*, pp. 243–244.

94. Diesel, *Solidarismus*, pp. 5–6, 31–34. Diesel provides detailed sample contracts in Part II.

95. Ibid., p. 24.

96. Ibid., p. 5.

97. Ibid., p. 10. Diesel even provides a sample card, on p. 105.

98. Ibid., pp. 16–18.

99. Braun, "Ingenieure und soziale Frage," p. 869, mentions the influence of the seventeenth- and eighteenth-century social contract theorists, especially Hobbes and Locke.

100. Leo A. Loubere, *Louis Blanc* (Evanston, Ill.: Northwestern University Press, 1961), pp. 34–39; and G. D. H. Cole, *A History of Socialist Thought*, Vol. II: *The Forerunners, 1789–1850* (New York: St. Martin's Press, 1953), pp. 66–71.

101. Cole, *History*, pp. 209–212; and George Woodcock, *Pierre Joseph Proudhon* (London: Routledge and Kegan Paul, 1956), pp. 122–123.

102. Diesel, *Solidarismus*, p. 30.

103. Ibid., pp. 5, 34, 43, 46, 52–53, 62.

104. Ibid., p. 61.

105. Ibid., p. 105; and Diesel, "Erläuternde Ergänzungen."

106. Diesel, "Erläuternde Ergänzungen."

107. Copies of these speeches are in the M.A.N. Werkarchiv.

108. Notes in the M.A.N. Werkarchiv.

109. Review by Dr. Arthur Mulberger, in *Schweiz. Konsum-Verein*, IV, No. 8, February 20, 1904, pp. 57–58. Diesel summarized the reviews from the *Sozialische Monatshefte*, Nr. 2, February 1904, and *Documente des Sozialismus*, IV, Nr. 1, pp. 6–7, in unpublished notes in the M.A.N. Werkarchiv.

110. Rudolf Diesel, "Denkschrift über den Stand der Diesel-Motor-Unternehmungen am 30. Juni 1898," in the Allgemeine Gesellschaft für Dieselmotoren *Nachlass*, M.A.N. Werkarchiv.

111. Kurt Schnauffer, "Aus der Entwicklung des Dieselmotors im In- und Ausland in den ersten zwanzig Jahren," *Motortechnische Zeitschrift*, 33 (1972), 81.

112. Review of *Solidarismus* by Karl Figdor, in *Monatsblätter des wissenschaftliche Klub in Wien*, December 30, 1904, p. 19. Unfortunately most of Oldenbourg's pre-1939 archival material was lost during World War II, so it is not possible to find more information on the publishing of the book. Oldenbourg Verlag to the author, July 6, 1984.

113. See Braun, "Ingenieure und soziale Frage," pp. 867–868, 870–874, for discussions of Reuleaux's, Bach's and Molendorff's organic theories. The characteristics of nineteenth-century utopias are discussed in Frank E. Manuel, "Toward a Psychological Theory of Utopias," in Frank

E. Manuel (ed.), *Utopias and Utopian Thought* (Boston: Houghton Mifflin, 1966), pp. 79–85.

114. Diesel, *Solidarismus*, pp. 29, 53.

115. The process of *Sammlung*, or the rallying of classes to maintain the old order, was analyzed in terms of Germany's naval fleet-building program by Eckart Kehr. See his *Battleship Building and Party Politics in Germany, 1894–1901*, trans. Pauline R. Anderson and Eugene N. Anderson (Chicago: University of Chicago Press, 1975). For a more recent discussion of *Sammlungspolitik* in imperial Germany, see the first three chapters of V. R. Berghahn, *Germany and the Approach of War in 1914* (New York: St. Martins, 1973). Some engineers even called for a "spiritual Sammlungspolitik." See Gerhard Zweckbronner, "Je besser der Techniker, desto einseitiger sein Blick? Probleme des technischen Fortschritts und Bildungsfragen in der Ingenieurerziehung im deutschen Kaiserreich," in Ulrich Troitzsch und Gabriele Wohlauf (eds.), *Technik-Geschichte: Historische Beiträge und neuere Ansätze* (Frankfurt/M: Suhrkamp, 1980), p. 348.

116. Rosalind H. Williams, "Solidarism: An Answer to Reagan Darwinism," *New York Times*, July 2, 1981, p. A19.

## Chapter 3

1. He used this term to describe what the Introduction to this volume labels the periods of invention and development.

2. Diesel, *Enstehung*, pp. 1, 151–152.

3. Lynwood Bryant, "Rudolf Diesel and His Rational Heat Engine," *Scientific American*, 221 (August 1969), 108.

4. For the early history of the internal combustion engine, see, for example, Bryan Donkin, *A Text-Book on Gas, Oil, and Air Engines; or, Internal Combustion Motors without Boiler*, (2nd ed.; London: Charles Griffin and Co., Ltd., 1896); C. Lyle Cummins, Jr., *Internal Fire* (Lake Oswego, Oreg.: Carnot Press, 1976); Friedrich Sass, *Geschichte des deutschen Verbrennungsmotorenbaues von 1860 bis 1918* (Berlin: Springer-Verlag, 1962), pp. 2–15; and Eugen Diesel, Gustav Goldbeck, and Friedrich Schildberger, *From Engines to Autos*, trans. Peter White (Chicago: Henry Regnery, 1960), pp. 1–35.

5. Lynwood Bryant, "The Silent Otto," *Technology and Culture*, 7

(Spring 1966), 186. Double-acting engines produce combustion at both ends of the piston. Some diesel engines later utilized the double-acting process. A typically enthusiastic tract on the new engines is Gustav Consentius, *Die Dampfkraft durch das Gaskraft ersetzt* (Leipzig: Ernst Schafer, 1860).

6. Lynwood Bryant has emphasized this point to the author. See also Donkin's *Textbook*, especially appendix D, pp. 434–464, which lists hundreds of patents taken out on engines or parts for them between 1884 and 1895.

7. For this trend, see D. S. L. Cardwell, *From Watt to Clausius: The Rise of Thermodynamics in the Early Industrial Age* (Ithaca, N.Y.: Cornell University Press, 1971); and Lynwood Bryant, "The Role of Thermodynamics in the Evolution of Heat Engines," *Technology and Culture*, 14 (April 1973), 152–165.

8. S. S. Wilson, "Sadi Carnot," *Scientific American*, 245 (August 1981), 134.

9. See the American edition, trans. R. H. Thurston (New York: American Society of Mechanical Engineers, 1943).

10. Wilson, "Sadi Carnot," pp. 134–138; and Cardwell, *Watt to Clausius*, p. 192.

11. Fragments of his notebooks indicate that he was in fact rejecting the caloric theory and moving toward the idea of the equivalence of heat and work. Wilson, "Sadi Carnot," p. 145.

12. Ibid., pp. 143–144; and Cardwell, *Watt to Clausius*, pp. 193, 207.

13. Stanley W. Angrist and Loren G. Hepler, *Order and Chaos: Laws of Energy and Entropy* (New York: Basic Books, 1967), p. 161.

14. Bryant, "Diesel and his Rational Engine," p. 111. See also D. S. L. Cardwell, *Turning Points in Western Technology* (New York: Neale Watson, 1972), pp. 131–133.

15. The Carnot cycle is described in Bryant, "Diesel and his Rational Engine," pp. 110–113; Wilson, "Sadi Carnot," pp. 140–141; Cardwell, *Watt to Clausius*, pp. 192–201; and Cardwell, *Turning Points*, pp. 131–133.

16. Angrist and Hepler, *Order and Chaos*, pp. 154, 156; and Malcolm W. Browne, "Perpetual Motion?," *New York Times*, June 8, 1985, C2.

17. Carnot, *Reflections*, p. 107.

18. Wilson, "Sadi Carnot," p. 144.

19. Paul Meyer, "Geschichte des Dieselmotors," pp. II.A. 9–12, manuscript, copies in the M.A.N. Werkarchiv and the Deutsches Museum.

20. Diesel to Zeuner, January 8, 1893, Diesel *Nachlass*, Deutsches Museum.

21. Meyer, "Geschichte des Dieselmotors," pp. II.A. 9 and IV.B. 48.

22. Bryant, "Role of Thermodynamics," p. 159. Modern steam cycle efficiencies range from 30 to 40 percent.

23. Dorette Breig, Diesel's granddaughter, to the author, March 7, 1979.

24. Bryant, "Role of Thermodynamics," p. 160.

25. Ibid. Lynwood Bryant has also emphasized these points to the author.

26. The story of the invention of the otto engine is told in several places. See, for example, Part 1 of Sass's *Geschichte*; Goldbeck, *Gebändigte Kraft*; Goldbeck, "Nicholas August Otto, Creator of the Internal-Combustion Engine," in Diesel, Goldbeck, and Schildberger, *From Engines to Autos*, pp. 36–84; and Lynwood Bryant, "The Silent Otto," pp. 184–200.

27. Bryant, "Silent Otto," pp. 191–192.

28. Ibid., p. 184.

29. Lynwood Bryant has argued, however, that Otto was really the first to use the four-stroke cycle in an actual internal combustion engine and should be given credit for its invention. See his article, "The Origin of the Four-Stroke Cycle," *Technology and Culture*, 8 (April 1967), 178–198.

30. Bryant, "Role of Thermodynamics," pp. 161–164. See also Donkin's *Textbook*.

31. Donkin, *Textbook*, pp. 2, 3, 9.

32. Ibid., p. 14.

33. Ibid., p. 224; and Bryant, "Role of Thermodynamics," pp. 164–165.

34. Copy of Diesel's list of courses at the Munich Technische Hochschule, in the M.A.N. Werkarchiv.

35. Diesel, *Entstehung*, pp. 1–2. Diesel probably inadvertently slipped into using the expression "available" when he meant total heat energy of the fuel.

36. Eugen Diesel, *Diesel*, p. 471.

37. Note from Diesel's "Theoretische Maschinenlehre II + III," winter semester 1878–79, in the M.A.N. Werkarchiv.

38. Eugen Diesel, *Diesel*, pp. 115–116.

39. From a manuscript fragment entitled "Aus Gracians Handorakel No. 231," in the M.A.N. Werkarchiv. This manuscript was apparently a response to attacks on Diesel's book *Die Entstehung des Dieselmotors*.

40. Eugen Diesel, *Diesel*, p. 137.

41. See chapter 1.

42. Diesel's unpublished manuscripts, "Abhandlung über den Ammoniak-Motor, System Diesel, 1880–1887/88," and "$NH_3$ Motor Memorandum," both in the M.A.N. Werkarchiv, record his plans and tests during the 1880s.

43. Diesel to Linde, February 11, 1892, Diesel *Nachlass*, Deutsches Museum.

44. The manuscript version uses the French "Construction." The printed version was entitled *Theorie und Konstruktion eines rationellen Wärmemotors zum Ersatz der Dampfmaschinen und der heute bekannten Verbrennungsmotoren* (Berlin: Julius Springer, 1893).

45. Diesel's correspondence with Springer Verlag, which eventually published his manuscript, throws no light on this question. Shortly after Springer Verlag agreed to publish the book, Diesel met with the publisher to iron out details of their agreement, and possibly at this meeting the decision was made not to print the "Concluding Remarks." Springer to Diesel, October 3, 1892. The Diesel-Springer correspondence was on loan from the publishing house to the Deutsches Museum in the summer of 1984. Lynwood Bryant has suggested the possibility that Diesel actually added the "Concluding Remarks" at a later time to justify his claims to originality. An examination of the original manuscript, in the Special Collections Department of the Deutsches Museum, does not reveal any obvious evidence of a later addition. Perhaps Diesel tampered with documents he knew would be read subsequently by historians, but the author remains unconvinced.

46. Diesel, "$NH_3$ Motor Memorandum," p. M1.

47. Bryant, "Role of Thermodynamics," pp. 161–162.

48. Diesel, "$NH_3$ Motor Memorandum," pp. M6–M7.

49. Schnauffer, "Die Erfindung des Dieselmotors, 1890–1893," p. 2, unpublished manuscript, copies in the M.A.N. Werkarchiv and the Spe-

cial Collections Department, Deutsches Museum. As a liquid is boiling and evaporating, its temperature will remain constant, depending on the liquid's boiling point and its pressure. Once it is completely evaporated, the vapor can be further heated, causing its temperature to rise and its volume to increase. This is the process of superheating.

50. A direct relationship exists between the pressure developed in a steam or internal combustion engine and the work done. The greater the pressure, the more powerful the engine. Hence, for a given set of engine dimensions, the higher the pressure realized in the cylinder, the higher the power. Limitations on the engine's increased power will, of course, be provided by such factors as internal frictions and the necessity for thicker cylinder walls.

51. Diesel, "Abhandlung," pp. 1–6, and "$NH_3$ Motor Memorandum," pp. 1–2. During this period, Diesel was in contact with Zeuner concerning his experiments with various solutions of ammonia, glycerin, and water. See Diesel, *Entstehung*, p. 152, note 2.

52. Diesel, "Abhandlung," p. 30.

53. Ibid., p. 49.

54. Ibid., p. 130.

55. See appendix 2 for the German original of this document.

56. Diesel, "Schlussbemerkungen," pp. 62–63, emphasis added.

57. Diesel, "Diesels rationeller Wärmemotor," p. 2.

58. Ibid. The atmosphere (atm) is a unit of pressure. One atmosphere, or standard atmospheric pressure, is equivalent to 14.696 pounds per square inch.

59. Diesel, "Schlussbemerkungen," p. 63, emphasis added.

60. Diesel, "Diesels rationeller Wärmemotor," p. 2.

61. Ibid., emphasis added.

62. Diesel, "Schlussbemerkungen," p. 63, emphasis added.

63. Ibid., pp. 63–64, emphasis added.

64. Lynwood Bryant has stressed this point, especially in relation to Diesel, to the author.

65. Zeuner quoted in Diesel's "Schlussbemerkungen," p. 61.

66. Diesel, "Schlussbemerkungen," p. 61, emphasis added.

67. Diesel's first patent draft, "Neue rationelle Wärmemotor," pp. 1–5; his published patent description, *Arbeitsverfahren und Ausführungsart für Verbrennungskraftmaschinen*, Patentschrift Nr. 67207, pp. 1–2; and the English translation of Diesel's 1893 book, *Theory and Construc-*

*tion of a Rational Heat Engine* (London and New York: Spon and Chamberlain, 1894), p. 1. The following description of Diesel's theory of combustion is drawn largely from these three works, plus his 1897 Kassel lecture. He did not clarify how his theory originated. He said in his 1913 *Die Entstehung des Dieselmotors* (p. 2) that he did not know how he conceived the idea of replacing ammonia with highly compressed and heated air, gradually introducing fuel into the air, and allowing the air to expand during combustion so that as much as possible of the heat produced was transformed into work. Perhaps he did not wish to acknowledge that he borrowed some of his ideas from other sources. The account presented in this chapter is based primarily on documents already cited above that were prepared during the critical period of the late 1880s and early 1890s.

68. Diesel always argued that compression ignition was not the defining characteristic of his engine, but only one of a number of conditions that made up his theory of combustion. Friedrich Sass has argued that compression ignition is indeed the defining characteristic of the diesel engine, but he is speaking of the present-day engine, not of Diesel's theory. Diesel may have wished to downplay compression ignition because he thought engines using it already existed, a fact that would endanger his patent. In the context of his 1892 theory, however, compression ignition is only one of several defining elements. See Diesel's *Entstehung*, pp. 3–5; and Sass, *Geschichte*, pp. 406–408.

69. Diesel's expression was "isothermal combustion." Today, the expression "isothermal expansion" would be substituted.

70. $\frac{T_{max} - T_{min}}{T_{max}} = \frac{1073 - 293}{1073} = .727$. See Diesel, *Theory and Construction*, pp. 43, 47–49; and Bryant, "Diesel and his Rational Engine," p. 115.

71. Bryant, "Diesel and his Rational Engine," pp. 113–114, and "The Development of the Diesel Engine," pp. 434–435.

72. Diesel, "Neue rationelle Wärmekraftmaschine," p. 8.

73. Ibid., p. 9.

74. Ibid., p. 10.

75. Diesel's compound engine is described in Diesel, *Theory and Construction*, pp. 54–63; Bryant, "Diesel and his Rational Engine," pp. 114–116; and Meyer, "Geschichte des Dieselmotors," pp. IV.B. 33–34.

76. Diesel, *Theory and Construction*, p. 60.

77. Diesel, "Schlussbemerkungen," pp. 62, 63. Diesel was referring

to Zeuner's comments on the combustion process, quoted earlier, and a comment from the French engine expert Gustave Richard to the effect that the phenomenon of combustion was a riddle and that the only immediate hope of improvement was through the use of a regenerator.

78. Schnauffer, "Die Erfindung des Dieselmotors," p. 6.

79. Leipzig: Baumgärtners Buchhandlung.

80. Köhler also described this engine in his review of Diesel's book *Theorie und Konstruktion* entitled "Die rationelle Wärmemotor im Vergleich mit anderen Wärmemotoren," *Zeitschrift des Vereins Deutscher Ingenieure*, 37 (September 9, 1893), 1106–1107.

81. Franz Grasshof, *Theoretische Maschinenlehre*, Vol. III: *Theorie der Kraftmaschinen* (Hamburg and Leipzig: Leopold Voss, 1890), pp. 886–891.

82. Bryant, "Development of the Diesel Engine," p. 435, note 6.

83. Eugen Diesel, *Diesel*, p. 474.

84. Meyer, "Geschichte des Dieselmotors," pp. IV.B. 5, 9.

85. Diesel, "Diesels rationeller Wärmemotor," p. 3, note 1.

86. Manuscript in the Diesel *Nachlass*, Deutsches Museum.

87. Sass, *Geschichte*, p. 414.

88. Cummins, *Internal Fire*, pp. 267, 277.

89. Diesel, "Schlussbemerkungen," p. 60.

90. Meyer, "Geschichte des Dieselmotors," pp. IV.B. 12–17.

91. Köhler, "Die rationelle Wärmemotor," p. 1107. Köhler had already made the same criticisms in his 1887 book after he had described the theory of a compound engine operating on the Carnot cycle. This is another reason to believe that Diesel was unaware of Köhler until 1893.

92. Schnauffer, "Die Erfindung des Dieselmotors," pp. 13–14.

93. Alfred Heggen, *Erfindungsschutz und Industrialisierung in Preussen, 1793–1877* (Göttingen: Vandenhoeck und Ruprecht, 1974), pp. 121–125, 142.

94. Heinrich Rohn, "Erfindungsbegriff im deutschen Patentgesetze," *VDI-Zeitschrift*, 37 (July 15, 1893), 843.

95. Goldbeck, *Gebändigte Kraft*, pp. 203–204.

96. F. Damme, *Das Deutsche Patentrecht: Ein Handbuch für Praxis und Studium* (2nd ed.; Berlin: Otto Liebemann, 1911), pp. 459–498, and "Übersicht über die Thätigkeit des deutschen Patentamtes seit 1877," *VDI-Zeitschrift*, 35 (February 21, 1891), 235.

97. Rohn, "Erfindungsbegriff," p. 843.

98. Riedler, *Dieselmotoren*, pp. 233–249.

99. Übersicht über die Thätigkeit des deutschen Patentamtes," pp. 234–236, and *Die Geschaftstätigkeit des kaiserlichen Patentamts und die Beziehung des Patentschutzes zu der Entwicklung der einzelnen Industriezweige Deutschlands in den Jahre 1891 bis 1900* (Berlin: Carl Heymann, 1902), pp. 328–329, 337.

100. Meyer, "Geschichte des Dieselmotors," pp. IV.A. 2.

101. Claims seven through twelve deal with such things as exhaust gases and the use of powdered coal as fuel. They tend to be somewhat repetitious of the first six. Claim twelve does discuss both single-cylinder and compound engines.

102. Schnauffer, "Erfindung des Dieselmotors," pp. 16–17.

103. The documents from the Patent Office and patent attorneys are no longer available. See Meyer, "Geschichte des Dieselmotors," p. IV.A. 1. The M.A.N. Werkarchiv does contain, however, a typed summary that includes excerpts of the various Patent Office decisions during the application stage.

104. Diesel's printed patent description, *Arbeitsverfahren und Ausführungsart für Verbrennungskraftmaschinen*, pp. 6–7.

105. Diesel, *Enstehung*, pp. 152–153.

106. Paul Meyer, "War der Dieselmotor jemals durch Patente geschützt?" *Schweizerische Bauzeitung*, May 28, 1949, pp. 309–310.

107. Georg Strössner, "50 Jahre Hauptpatent Rudolf Diesels," *Motortechnische Zeitschrift*, 2 (1942), 37.

108. Schnauffer, "Die Erfindung Rudolf Diesels," pp. 22–24; and Sass, *Geschichte*, pp. 391–393.

109. Sass, *Geschichte*, pp. 392–393.

110. E. D. Meier, "Report on the Diesel Engine to Adolphus Busch," p. 2, from a copy made in 1912, in the M.A.N. Werkarchiv.

111. Letter to Schroeter, April 13, 1892, copy in the M.A.N. Werkarchiv.

112. Sass, *Geschichte*, p. 393.

113. Schnauffer, "Die Erfindung des Dieselmotors," p. 25.

114. Langen to Diesel, April 19, 1892, in the Diesel *Nachlass*, Deutsches Museum. See also chapter 4.

115. Meyer, "Geschichte des Dieselmotors," pp. IV.B. 2–6, quote is on p. IV.B. 3; and examination of the Diesel-Springer correspondence.

116. Diesel to M.F.A. (Augsburg Engine Works), October 12, 1892, Diesel *Nachlass*, Deutsches Museum.

117. Meyer, "Geschichte des Dieselmotors," pp. V.B. 5; and Sass, *Geschichte*, p. 394.

118. Diesel to his wife, July 21, 1893, M.A.N. Werkarchiv.

119. On Donkin, see the *Dictionary of National Biography*, Second Supplement (London: Smith, Elder and Co., 1912), I, 513–514. Diesel approached Springer several times requesting suggestions for an English translator. Springer may have suggested Donkin, but no evidence of this exists in their correspondence.

120. Schnauffer, "Die Erfindung des Dieselmotors," pp. 26–30. A typical letter was that to Siemens and Halske of January 19, 1893, M.A.N. Werkarchiv.

121. Sass, *Geschichte*, p. 395; and Schnauffer, "Die Erfindung des Dieselmotors," pp. 32–34.

122. Schroeter's appeared on February 4, 1893, in the *Bayerisches Industrie und Gewerbeblatt*, and Gutermuth's on March 11, 1893, in the *VDI-Zeitschrift*.

123. Moritz Schroeter, "Ein neuer Wärmemotor," Separat-Abdruck aus *Bayerisches Industrie- und Gewerbeblatt*, No. 5, February 4, 1893, copy in M.A.N. Werkarchiv.

124. Paul Meyer to Erich von Kurzel-Runtscheiner, January 19, 1941, M.A.N. Werkarchiv.

125. Langen to Diesel, January 13, 1893, Diesel *Nachlass*, Deutsches Museum.

126. Sass, *Geschichte*, pp. 397–398.

127. Schnauffer, "Die Erfindung des Dieselmotors," pp. 41, 42, 47–48.

128. The manuscript is in the Diesel *Nachlass*, Deutsches Museum.

129. Diesel, "Nachträge," p. 1. In his 1913 *Entstehung*, Diesel finally admitted that practical difficulties with the realization of isothermal combustion and with mechanical efficiency made him abandon constant-temperature combustion and adopt the idea of constant-pressure combustion. He defends not publishing these documents at the time by saying it was not in the interest of industry to reveal information about the new engine it was developing. See pp. 3, 153–154, note 5. Because his patents had expired, Diesel could afford to be more informative about the evolution of his theory into practice. Unfortunately, it was too late to stop the attacks on him, which had been going on for some years.

130. Diesel, "Nachträge," p. 192; and Schnauffer, "Die Erfindung des Dieselmotors," p. 42.

131. Diesel, "Nachträge," pp. 192–193.
132. Ibid.; and Schnauffer, "Die Erfindung des Dieselmotors," p. 43.
133. Diesel, "Nachträge," p. 171, emphasis added.
134. Schnauffer, "Die Erfindung des Dieselmotors," p. 43.
135. Ibid.
136. Ibid., p. 46; and Diesel, "Nachträge," p. 211.
137. Schnauffer, "Die Erfindung des Dieselmotors," pp. 43–44.
138. Sass, *Geschichte*, pp. 404–405.
139. Ibid., pp. 405–406; and Schnauffer, "Die Erfindung des Dieselmotors," pp. 45, 61–62.
140. Schnauffer, "Die Erfindung des Dieselmotors," p. 63.
141. Ibid., p. 47; and Sass, *Geschichte*, p. 405.
142. Letters to Buz and Krupp, both dated September 26, 1893, Diesel *Nachlass*, Deutsches Museum.
143. Schnauffer, "Die Erfindung Rudolf Diesels," p. 319.
144. Schnauffer, "Die Erfindung des Dieselmotors," p. 56.
145. This discussion of Diesel's second patent is largely drawn from Schnauffer, "Die Erfindung des Dieselmotors," pp. 65–69; and Sass, *Geschichte*, pp. 408–412.
146. Schnauffer, "Die Erfindung des Dieselmotors," p. 66.
147. Ibid., p. 68.
148. See also Meyer, "War der Dieselmotor jemals durch Patente geschützt?" p. 309.
149. Lynwood Bryant has emphasized this point to the author.
150. Letter in M.A.N. Werkarchiv.
151. Schnauffer, "Die Erfindung des Dieselmotors," p. 69. In his *Entstehung* (pp. 3 and 152, note 3), Diesel still made it seem as though constant-pressure combustion was achieved by manipulating the injection time of the fuel. He did not answer attacks on his second patent but said, because the patent was never invalidated, it was useless to speculate about it so many years later.

## Chapter 4

1. Schnauffer, "Die Erfindung des Dieselmotors," p. 24.
2. See, for example, Diesel's first letter to Krupp, January 19, 1893, in the Diesel *Nachlass*, Deutsches Museum.
3. Linde, *Aus meinem Leben*, pp. 39–48; H. Droscha, "Geschichte

der M.A.N.," p. 20, unpublished manuscript in the M.A.N. Werkarchiv; Friedrich Klemm, "Carl von Linde," in *Dictionary of Scientific Biography*, VIII, 356–366.

4. Letter of April 16, 1892, in M.A.N. Werkarchiv; and Kurt Schnauffer, "Diesels erste Verträge," pp. 4–5, unpublished manuscript, copies in the M.A.N. Werkarchiv and the Special Collections Department, Deutsches Museum.

5. Schnauffer says Linde took great pains to aid Diesel, but he offers no documentation.

6. Schnauffer, "Diesels erste Verträge," p. 13.

7. Heinrich Bechtel, *Wirtschafts-und Sozialgeschichte Deutschlands* (Munich: Georg D. W. Callweg, 1967), pp. 378–379.

8. Hans Rosenberg, "Wirtschaftskonjunktur, Gesellschaft, und Politik in Mitteleuropa, 1873 bis 1896," in Hans Ulrich-Wehler (ed.), *Moderne Deutsche Sozialgeschichte* (Cologne-Berlin: Kiepenheuer und Witsch, 1966), p. 236.

9. Droscha, "Geschichte der M.A.N.," p. 20; and "Notizen zu den [M.F.A.] Aufsichtsrathsitzungen, 19. Juli 1890–21. Okt. 1898," M.A.N. Werkarchiv.

10. Georg Strössner, "Die Fusion der Aktiengesellschaft Maschinenfabrik Augsburg und der Maschinenbau-Actien-Gesellschaft Nürnberg im Jahre 1898," *Tradition: Zeitschrift für Firmen-Geschichte und Unternehmerbiographie*, V (June 1960), 115.

11. Josef Krumper, "Einige Lebenserinnerungen," 1916, p. 53, unpublished manuscript in the M.A.N. Werkarchiv.

12. This was certainly Krupp's judgment. See report of Krupp directors Asthöwer and Klüpfel to F. A. Krupp of March 18, 1893, in Wilhelm Worsoe, "Die Mitarbeit der Werke Fried. Krupp an der Entstehung des Dieselmotors in den Jahren 1893/97 und an der Anfangs-Entwicklung in den Jahren 1897/99," Kiel: n.p., 1933/1940, Anlage 8, manuscript in the Diesel *Nachlass*, Deutsches Museum. A copy is also in the M.A.N. Werkarchiv.

13. Schnauffer, "Diesels erste Verträge," p. 2; and Sass, *Geschichte*, p. 425.

14. Letter to Diesel of April 2, 1892, in Diesel *Nachlass*, Deutsches Museum.

15. Langen to Diesel, April 19, 1892, Diesel *Nachlass*, Deutsches Museum.

16. Schnauffer, "Diesels erste Verträge," p. 14.

17. Diesel to Buz, April 6, 1892, Diesel *Nachlass*, Deutsches Museum. This period is also described in Sass, *Geschichte*, pp. 426–427; and Schnauffer, "Diesels erste Verträge," pp. 6–13.

18. Schnauffer, "Diesels erste Verträge," pp. 8–9.

19. Diesel to Buz, April 9, 1892, Diesel *Nachlass*, Deutsches Museum.

20. Diesel to Schroeter, April 13, 1892, M.A.N. Werkarchiv.

21. Buz to Diesel, April 20, 1892, Diesel *Nachlass*, Deutsches Museum.

22. Diesel to Linde, April 6, 1892, Diesel *Nachlass*, Deutsches Museum.

23. Kurt Schnauffer is of this opinion. See his "Diesels erste Verträge," p. 15.

24. Strössner, "Die Fusion Maschinenfabrik Augsburg und Maschinenbau Nürnberg," p. 100.

25. Letter in M.A.N. Werkarchiv.

26. See Diesel, *Entstehung*, p. 7. The first letter in the Diesel-Augsburg correspondence in which Diesel uses the "Du" form with Vogel is, so far as the author can determine, June 13, 1894. Vogel responded in kind on June 14. Letters in the Diesel *Nachlass*, Deutsches Museum.

27. Krumper to Buz, November 13, 1913, M.A.N. Werkarchiv.

28. Krumper, "Einige Lebenserinnerungen," pp. 59–60.

29. Buz to Krumper, December 1, 1913, M.A.N. Werkarchiv, emphasis added. In his memoirs, Krumper also speaks of a letter of December 1913 from Lucian Vogel that further supports his side of the story. The author was unable to locate such a letter in the M.A.N. Werkarchiv.

30. Letters in Diesel *Nachlass*, Deutsches Museum.

31. Immanuel Lauster, "Der Dieselmotor," pp. 53–54, 85–86, unpublished manuscript from the 1930s and early 1940s in the M.A.N. Werkarchiv. In 1904 Lauster became the head of the diesel engine division. Later, he became a member of the M.A.N. board of directors and for several years in the 1930s was its chairman. He died in 1948. See Sass, *Geschichte*, p. 521.

32. Diesel to Buz, July 1, 1913, M.A.N. Werkarchiv.

33. Meier, "Report," pp. 6, 10.

34. Letter in Diesel *Nachlass*, Deutsches Museum.

35. Sass, *Geschichte*, p. 395.

36. Letter in Diesel *Nachlass*, Deutsches Museum.

37. Letter of January 13, 1893, Diesel *Nachlass*, Deutsches Museum.

38. Treue, *Eugen Langen und Nic. August Otto*, pp. 91–93. Goldbeck says that, though it cannot be proven, Otto Köhler may have been advising Langen to reject Diesel's approaches. See his *Gebändigte Kraft*, p. 242. As was seen in chapter 3, Köhler had criticized Diesel's ideas in 1893 and said the attempt to realize the Carnot cycle was bound to fail. In 1897 Köhler cooperated with Deutz when they threatened to bring suit against the validity of Diesel's patent. See chapter 5.

39. Letter of June 1, 1893, Diesel *Nachlass*, Deutsches Museum. Crossley's attitude may have been influenced by the fact that it was an English licensee of Deutz.

40. Letter in Diesel *Nachlass*, Deutsches Museum.

41. Letter in Diesel *Nachlass*, Deutsches Museum.

42. Report of Krupp directorate to F. A. Krupp, February 4, 1893, in Worsoe, "Die Mitarbeit Krupp," Anlage 2. Krupp had recently taken over the Grusonwerk, in Magdeburg, which was partly involved in building gas engines. Krupp apparently hoped that Diesel's process could be adapted to large gas engines. Kurt Schnauffer, "Anteil Krupps an der Entwicklung des Dieselmotors," p. 1, unpublished manuscript in M.A.N. Werkarchiv.

43. Worsoe, "Die Mitarbeit Krupp," Anlage 3. A marginal note by F. A. Krupp indicates complete agreement with the directors' conclusions.

44. Diesel once again was stretching the truth. Langen was unreceptive to his proposals, and, though the two met on January 29, later letters provide no evidence that Langen was at all prepared to support Diesel. Deutz did not become interested in the diesel engine until 1896, after it had been successfully developed.

45. Diesel to Krupp, February 18, 1893, Diesel *Nachlass*, Deutsches Museum.

46. Worsoe, "Die Mitarbeit Krupp," Anlagen 6, 10, and 11; and Schnauffer, "Diesels erste Verträge," pp. 16–17, 21–25.

47. Schnauffer, "Diesels erste Verträge," pp. 16, 17, 21–25.

48. A Linde Company information sheet dated September 1, 1893, announced Diesel's resignation. Copy in Diesel *Nachlass*, Deutsches Museum. Diesel continued to advise the Linde Company about his activities, presumably to keep open the possibility of some use of diesel engines for refrigeration.

49. Schnauffer, "Diesels erste Verträge," p. 17. By 1897 Augsburg and

Krupp had spent 175,455 marks on the tests. In addition, Krupp had spent 120,000 marks on Diesel's salary during the four-year testing period. From a M.A.N. report prepared during the 1960s as a source for Kurt Schnauffer's Diesel manuscript in the M.A.N. Werkarchiv.

50. Worsoe, "Die Mitarbeit Krupp," p. 7.

51. Diesel to Martha, February 13, 1898, M.A.N. Werkarchiv.

52. Worsoe, "Die Mitarbeit Krupp," p. 7.

53. Schnauffer, "Die Entwicklung des Dieselmotors bei Fried. Krupp, Essen," pp. 4–6, unpublished manuscript in the M.A.N. Werkarchiv.

54. Diesel to Martha, October 8, 1896, M.A.N. Werkarchiv.

55. For example, Diesel to Martha, May 23, 1895, ibid.

56. Diesel to Martha, June 22, 1894, ibid.

57. See for example, Diesel's Entstehung, pp. 78–79.

58. In preparation for his speech and book, Diesel reviewed relevant material and wrote former colleagues asking them to describe their part in the process. A number of these letters are in the Eugen Diesel Nachlass, Freiburg. Friedrich Sass's Geschichte (pp. 431–478) thoroughly discusses the development of the engine. Unless otherwise indicated, the following summary is drawn from the works of Diesel and Sass. Lynwood Bryant presents an interesting discussion of the main problems during this period in his "The Development of the Diesel Engine," 432–446.

59. For example, letters of July 25, August 3 and 11, 1893, in M.A.N. Werkarchiv.

60. Diesel, Entstehung, p. 13.

61. Ibid., p. 14.

62. Diesel to Martha, August 12, 1893, M.A.N. Werkarchiv.

63. Ibid.

64. Diesel, Entstehung, p. 15.

65. Bryant, "Development of the Diesel Engine," p. 438.

66. Ibid., pp. 438–439; and Diesel, Entstehung, p. 13.

67. Diesel, Entstehung, pp. 23, 57, 59.

68. Diesel to Martha, March 6, 1894, M.A.N. Werkarchiv.

69. Lynwood Bryant has emphasized this point to the author.

70. Diesel to Martha, October 3, 1894, M.A.N. Werkarchiv.

71. Diesel to Martha, October 14, 1894, ibid.

72. From the entry on Reichenbach in a folder on "Leading Personalities" connected with the development of the diesel engine, in M.A.N. Werkarchiv.

73. A dynamometer measures power output.

74. Diesel, *Entstehung*, pp. 45–47. Effective horsepower is measured at the output of the system, after mechanical losses from such things as friction have been subtracted. Indicated horsepower, which was also measured in these engine tests, is measured in the cylinder before mechanical losses are subtracted.

75. Diesel to Martha, June 30, 1895, M.A.N. Werkarchiv.

76. Diesel to Martha, July 11, 1895, ibid.

77. Worsoe, "Die Mitarbeit Krupp," Anlage 33; and Meyer, "Geschichte des Dieselmotors," p. X. A.-C. 1.

78. Kurt Schnauffer, "Umbauten des Versuchmotors," 59–63, manuscript in the M.A.N. Werkarchiv; and Sass, *Geschichte*, pp. 464–465.

79. For a discussion of oil tariffs and their effect on the sale of diesel engines, see chapter 5.

80. Diesel to Martha, October 27, 1895, M.A.N. Werkarchiv.

81. Krupp to Diesel, November 6, 1895, and Diesel to Krupp, November 10, 1895, Diesel *Nachlass*, Deutsches Museum.

82. M.F.A. to Krupp, September 16, 1896, which quotes the letter of January 23, M.A.N. Werkarchiv.

83. Diesel to Krupp, January 23, 1896, M.A.N. Werkarchiv.

84. Both letters in ibid.

85. Ebbs to Krupp, January 30, 1896, in Worsoe, "Die Mitarbeit Krupp," Anlage 37.

86. Diesel, *Entstehung*, p. 59.

87. Ibid., p. 74.

88. Diesel to Buz, February 10, 1897, M.A.N. Werkarchiv.

## Chapter 5

1. Diesel to Martha, January 28, 1897, M.A.N. Werkarchiv.

2. Diesel had met Dyckhoff in 1882 in connection with Linde Company business. They became good friends, and Dyckhoff was one of the diesel engine's early supporters.

3. Quoted in E. D. Meier, "The Diesel Motor," *Engineering News*, 39, No. 11 (March 17, 1898), 172–174.

4. Diesel, "Diesels rationeller Wärmemotor," p. 8.

5. Schroeter, "Diesels rationeller Wärmemotor," p. 13.

6. Ibid., p. 17.

7. Ibid., p. 19.

8. Meier, "Report," p. 11. Other authorities have been puzzled by the contradiction in Schroeter's claims and the true status of the engine, but do not offer explanations for it. See, for example, Lauster, "Der Dieselmotor," p. 254.

9. Letters in Krupp-M.F.A. correspondence, M.A.N. Werkarchiv.

10. Diesel to Martha, March 12, 1896, ibid.

11. Contract of March 11, 1897, in Worsoe, "Die Mitarbeit Krupp," Anlage 46.

12. Kurt Schnauffer, "Die Motorentwicklung in Werk Augsburg der M.A.N., 1898–1918," p. 3, unpublished manuscript, copies in the M.A.N. Werkarchiv and the Deutsches Museum.

13. Kurt Schnauffer, "Lizenzverträge und Erstentwicklungen des Dieselmotors in In-und Ausland, 1893–1909," manuscript in the M.A.N. Werkarchiv. Anlagen Vol., p. 1, lists the total licensing agreements for each country; pp. 2–6 lists chronologically all German and foreign licensing agreements from 1893 to 1909.

14. The Augsburg-Nuremberg combine was not officially known as M.A.N. until 1908, but this abbreviation is used for simplicity's sake in the rest of this book for the years after 1898.

15. Sass, *Geschichte*, p. 482.

16. Schnauffer, "Lizenzverträge," pp. 8–14; and Sass, *Geschichte*, pp. 482–483, 490–491.

17. By 1901 the stock, then held by the Allgemeine, was all but worthless. Schnauffer, "Lizenzverträge," pp. 45–50.

18. Meier, "Report," p. 39.

19. Letters to Martha of March 3, 17, 20, and 23, 1897, in M.A.N. Werkarchiv.

20. Diesel to Martha, February 16, 1898, ibid.; and Eugen Diesel, *Diesel*, pp. 290–291.

21. Busch's biography, appearance, and life-style while in Germany are discussed in Eugen Diesel, *Diesel*, pp. 278–284; Eugen Diesel and Georg Strössner, *Kampf um eine Maschine: Die ersten Dieselmotoren in Amerika* (Berlin: Erich Schmidt Verlag, 1950), pp. 35–36; and Richard H. Lytle, "The Introduction of Diesel Power in the United States, 1897–1912," *Business History Review*, 42 (Summer 1968), 116.

22. Diesel and Strössner, *Kampf um eine Maschine,* pp. 36–38; and Schnauffer, "Lizenzverträge," pp. 63–71.

23. Stefan Loewengart, *From the History of My Family, The Bing Family of Nuremberg* (n.p., n.d.), pp. 14–23, copy in M.A.N. Werkarchiv.

24. Lytle, "Introduction of Diesel Power," pp. 119–148.

25. Meier, "Report," p. 2.

26. Ibid., p. 11.

27. Ibid., p. 15.

28. Ibid., pp. 14, 28.

29. Ibid., pp. 16–17.

30. Ibid., p. 28.

31. Ibid., p. 29.

32. "Lucian Vogel und seine Mitarbeit am Dieselmotor," manuscript dated 12 August 1940, and information from folder on "Leading Personalities" connected with the development of the diesel engine, both in M.A.N. Werkarchiv.

33. See Diesel and Strössner, *Kampf um eine Maschine,* pp. 52–55. A brochure prepared by the Diesel Motor Company for the exhibition claimed the combustion line of the indicator diagram was practically isothermal. The brochure promised that what German theory had produced, American skill would adapt to all uses. Copy in Diesel *Nachlass,* Deutsches Museum.

34. See Eugen Diesel, *Diesel,* pp. 305–312. The exhibition was also described by the two men whom Diesel put in charge of the diesel pavilion. See Meyer, "Geschichte des Dieselmotors," pp. IX.B. 1–4; and letters of Ludwig Noé to Rudolf Diesel of September 24, 1912, and to Eugen Diesel of March 13, 1940, both in the Eugen Diesel *Nachlass,* Freiburg.

35. Noé to E. Diesel, May 13, 1940. Diesel's illness was referred to as "neurasthenia cerebralis" in a letter of his patent lawyer F. C. Glaser, to the General Society, April 15, 1898, M.A.N. Werkarchiv. *Webster's Third International Dictionary* (1967) defines neurasthenia as a syndrome marked by fatigability of body and mind as well as usually by worry and depression and often accompanied by headaches and other disturbances. If Diesel was indeed a manic-depressive personality, this collapse was undoubtedly associated with the depressive stage. Busch quipped in a letter early in 1899 that he was not sure whether Diesel's illness came from overexertion or from having too much money. If the latter, he would

be glad to relieve Diesel of some of the 1,000,000 marks paid for the American patent. Adolphus Busch to the Allgemeine Gesellschaft, 1899, in the M.A.N. Werkarchiv.

36. Eugen Diesel, *Diesel*, pp. 318–322.

37. Sass, *Geschichte*, p. 486.

38. Krupp to M.F.A., March 2, 1897, M.A.N. Werkarchiv.

39. Krupp to Köhler, July 30, 1897, and Köhler to Krupp, August 3, 1897, M.A.N. Werkarchiv. See also Eugen Diesel, *Diesel*, pp. 268–269.

40. Cummins, *Internal Fire*, pp. 266–267.

41. Donkin, *Textbook*, p. 380. See also Cummins, *Internal Fire*, pp. 288–294; and Sass, *Geschichte*, pp. 220–224, 412–413, 486–488.

42. A copy of his original suit is in the M.A.N. Werkarchiv.

43. F. C. Glaser to Diesel, April 16, 1898, ibid.

44. Emil Capitaine, *Kritik des Dieselmotors* (Frankfurt: n.p., 1898), pp. 2–3, ibid.

45. The Patent Office's rejection of Capitaine's suit is reproduced in Rudolf Diesel, *Antwort auf Emil Capitaines Kritik des Dieselmotors* (Munich: n.p., 1898). Diesel contented himself with reprinting the court's decision and not answering specific criticisms.

46. A copy of the appeal is in the M.A.N. Werkarchiv.

47. Capitaine specifically denied that the request for an out-of-court settlement originated with Diesel. Capitaine to Justizrat Rosenthal, in Munich, May 29, 1899, ibid.

48. Krupp to M.F.A., July 7, 1898, ibid.

49. From a copy of the settlement in ibid.

50. Interestingly enough, the Diesel-Capitaine controversy was carried on into the next generation. In November 1937 Capitaine's son, Philipp, threatened legal action against Eugen Diesel and the publisher of his recent biography of his father, the Hanseatische Verlag in Hamburg, unless passages in the biography deemed detrimental to Emil Capitaine's reputation were removed or altered. Although both author and publisher responded that all passages were based strictly on documentation, this controversy also ended in a compromise. In February 1938 Eugen Diesel agreed to tone down certain passages about Capitaine in future editions. This, in fact, was done. Cf., for example, the comments on Capitaine on p. 294 of the 1937 edition with the same passage on p. 245 of the 1948 edition. Philipp wrote again in 1939 about the characterization of his father in the German film, *Diesel* (which would open in 1943). Eugen

worked on the film script and assured Philipp that his father would be treated fairly, and there the matter seems to have ended. The entire correspondence is contained in the Eugen Diesel *Nachlass*, Freiburg.

51. Sass, *Geschichte*, pp. 505–511; quote is on p. 511. The following discussion of the diesel engine's early difficulties is drawn from these pages as well as pp. 489–492. See also Kurt Schnauffer, "Aus der Entwicklung des Dieselmotors im In- und Ausland in den ersten zwanzig Jahren," *Motortechnische Zeitschrift*, 33 (1972), 81; and Schnauffer, "Die Motorentwicklung in Werk Augsburg der M.A.N., 1898–1918," pp. 18–20.

52. Eduard Blümel to Eugen Diesel, June 16, 1940, M.A.N. Werkarchiv.

53. Lauster, "Der Dieselmotor," p. 338.

54. Sass, *Geschichte*, p. 490.

55. Schnauffer, "Die Motorentwicklung in Werk Augsburg der M.A.N., 1898–1918," pp. 13–14; Sass, *Geschichte*, pp. 493–502; Bryant, "Development of the Diesel Engine," pp. 443.

56. Christian to Rudolf Diesel, October 30, 1897, M.A.N. Werkarchiv.

57. Diesel to Christian, October 31, 1897, ibid.

58. Christian to Diesel, November 8, 1897, and Gerstle to Diesel, November 8, 1897, ibid.

59. Schnauffer, "Diesels erste Verträge," p. 20.

60. Diesel to Bonnet and Co., November 18, 1897, M.A.N. Werkarchiv.

61. Christian referred to the first contact as "a banker, who although Jewish is very rich and acceptable." Christian to Rudolf Diesel, October 30, 1897, M.A.N. Werkarchiv.

62. Lynwood Bryant has emphasized this point to the author.

63. Buz to Diesel, November 19, 1897, M.A.N. Werkarchiv. In fact, extra capital in the form of an increased stock issue was voted. See Strössner, "Die Fusion Maschinenfabrik Augsburg und Maschinenbau Nürnberg," p. 101.

64. Diesel to Buz, November 20, 1897, M.A.N. Werkarchiv.

65. Christian, in a curious letter to Diesel, November 29, 1897, indicated that some tension existed between Bonnet and himself. Bonnet apparently did not want Christian to receive as much of the founding stock as he was at first supposed to receive and also did not want Christian on

the board of directors. Further, Christian had borrowed some kind of plans dealing with the new company and was suspected of copying them for his own purposes. In the letter, he expressed his innocence to Diesel and asked for his support in the stock issue, but said he would stay off the board if this was desired. Christian's role in the founding of the company is vague. Perhaps he began as a middleman between Diesel and the bankers, and then maneuvered his way onto the board. In a letter to M.A.N. of April 10, 1936, in the M.A.N. Werkarchiv, he says it was Diesel who invited him onto the board, but he is otherwise uninformative about the founding of the factory.

66. Diesel to Martha, undated, M.A.N. Werkarchiv.

67. Karl Dieterichs to Eugen Diesel, June 14, 1937, Eugen Diesel *Nachlass*, Freiburg; and Obering. Jansen, "Beitrag zur Geschichte des Dieselmotors," 1913, manuscript in the M.A.N. Werkarchiv; Lauster, "Der Dieselmotor," p. 53.

68. Letter of January 20, 1898, in M.A.N. Werkarchiv.

69. Dieterichs letter.

70. Information on the founding of the company is taken from Schnauffer, "Lizenzverträge, pp. 19–21; and "Auszug aus dem Allgemeinen Register Band VI, p. 45," manuscript in Allgemeine *Nachlass*, M.A.N. Werkarchiv.

71. Transcript of the *Augsburger Abendzeitung*, May 15, 1900, ibid.

72. List in the M.A.N. Werkarchiv.

73. Dieterichs letter.

74. Schnauffer, "Lizenzverträge," p. 22.

75. Letter of April 15, 1899, in Allgemeine *Nachlass*, M.A.N. Werkarchiv.

76. See for example, Ludwig Noé to Eugen Diesel, May 13, 1940, and Noé to Rudolf Diesel, September 24, 1912, Eugen Diesel *Nachlass*, Freiburg; and Dieterichs letter.

77. Memo in Allgemeine *Nachlass*, M.A.N. Werkarchiv.

78. Letter of September 25, 1899, ibid.

79. Ibid.

80. Schnauffer, "Lizenzverträge," pp. 30–31.

81. Letter of December 5, 1899, in Allgemeine *Nachlass*, M.A.N. Werkarchiv.

82. Letter of December 27, 1899, ibid.

83. Transcript of the *Augsburger Abendzeitung*, May 15, 1900, ibid.

84. Report in ibid.

85. Letter of May 15, 1900, ibid.

86. Noé to Eugen Diesel, May 13, 1940, Eugen Diesel *Nachlass*, Freiburg.

87. Transcript of the *Augsburger Abendzeitung*, May 27, 1900, M.A.N. Werkarchiv.

88. Meyer, "Geschichte des Dieselmotors," p. X.F. 8.

89. Lauster, "Der Dieselmotor," pp. 396–399.

90. Both letters quoted in Sass, *Geschichte*, pp. 485–486.

91. From a transcript in the M.A.N. Werkarchiv. Eugen Diesel calls this newspaper a "gossip paper." *Diesel*, p. 324.

92. Meyer, "Geschichte des Dieselmotors," pp. X.F. 6–8.

93. Lauster, "Der Dieselmotor," pp. 317, 355; and "Verhalten des Herrn Diesel gegenüber M.A.N.," M.A.N. Werkarchiv. The latter document was prepared during a time of ruptured relations between Augsburg and Diesel and needs to be treated with caution.

94. See, for example, the critical discussion following Diesel's lecture to the German Society of Naval Architects in November 1912, in *Jahrbuch der Schiffbautechnischen Gesellschaft*, 14 (1913), especially 359.

95. Meyer, "Geschichte des Dieselmotors," p. X.F. 6, so states. The author has seen no conclusive evidence on this point.

96. Report of the Allgemeine Gesellschaft, summer 1900, in the Allgemeine *Nachlass*, M.A.N. Werkarchiv.

97. Paul Meyer to Erich von Kurzel-Runtscheiner, January 19, 1941, M.A.N. Werkarchiv.

98. Lauster, "Der Dieselmotor," p. 330.

99. Jansen, "Beitrag zur Geschichte des Dieselmotors," p. 27.

100. The founding of the General Society is covered in Schnauffer, "Lizenzverträge," pp. 111–115. See also Sass, *Geschichte*, pp. 492–493.

101. Diesel to Bing, July 8, 1898, M.A.N. Werkarchiv.

102. Diesel to Buz, July 16, 1898, ibid.

103. Diesel to Martha, September 5, 1898, ibid. Diesel's fears concerning his health were soon borne out. In a letter of November 28, 1898, he informed Buz that he could not attend an upcoming meeting of the newly created society's board of directors because of his health and the necessity of entering a sanitarium. Letter in Allgemeine *Nachlass*, M.A.N. Werkarchiv.

104. In ibid.

105. Schnauffer, "Lizenzverträge," p. 114.
106. From a list of preferred stockholders of March 15, 1900, in Allgemeine Nachlass, M.A.N. Werkarchiv.
107. Schnauffer, "Lizenzverträge," p. 115; and Bing to Buz, July 11, 1898, in Allgemeine Nachlass, M.A.N. Werkarchiv.
108. In 1901 the General Society moved its offices to M.A.N. to lower costs. Worsoe, "Die Mitarbeit Krupp," p. 64.
109. Copies of these circulars are in the Diesel Nachlass, Deutsches Museum.
110. Proposal of October 1, 1900, in Allgemeine Nachlass, M.A.N. Werkarchiv.
111. See for example, "Streifzüge durch den Dieselmotor-Bau," May 1901, and the General Society's business report dated June 11, 1902, ibid.
112. Nuremberg: Wilh. Tümmel, 1901, copy in the Diesel Nachlass, Deutsches Museum.
113. Diesel to Felix Deutsch, director of the A.E.G., March 15, 1901, M.A.N. Werkarchiv.
114. "Streifzüge," cited above.
115. The Diesel Engine—The Rational Engine—Source of Power of the Future, Diesel Motor Company of America, May 1898, in Diesel Nachlass, Deutsches Museum.
116. Buz to Diesel, November 19, 1897, M.A.N. Werkarchiv.
117. Diesel Company of America, The Diesel Engine, p. 4.
118. Other brochures examined, e.g., from Deutz, Augsburg, and Nuremberg, emphasize the same features listed above.
119. Sass, Geschichte, p. 529.
120. Reports on June 28, 1901, and June 11, 1902, in Allgemeine Nachlass, M.A.N. Werkarchiv.
121. Diesel to Buz, July 23, 1899, ibid.
122. Diesel to Buz, April 21, 27, and 29, 1899, ibid., discuss this proposed transaction.
123. Worsoe, "Die Mitarbeit Krupp," Anlage 77.
124. According to Sass, Geschichte, p. 529, the oil tariff was reduced to 3.60 marks per 100 kilograms in 1906 and to 1.80 marks in 1912.
125. Ibid.; and Fred S. Baumann, Das Erdöl in Deutschland (Berlin: Carl Heymann, 1930), pp. 44–45.
126. Statistics taken from the General Society business reports of June 28, 1901, and June 11, 1902, in Allgemeine Nachlass, M.A.N. Werk-

archiv; and from lists of engines in production and use, February 1899 and September 1901, in Diesel *Nachlass*, Deutsches Museum.

127. From *Statistik des Deutschen Reiches*, Neue Folge, Band 113 (Berlin, 1898), reprinted in Goldbeck, *Gebändigte Kraft*, pp. 115–116.

128. For example, a letter of Diesel to the Allgemeine of January 15, 1900, in which he indicated that it could not transfer possession of his patents to foreign parties, only the right to use such patents. Allgemeine *Nachlass*, M.A.N. Werkarchiv.

129. Excerpts from Allgemeine Register, transcript, ibid.

130. Diesel to Johanning, July 4, 1901, ibid.

131. Allgemeine Register excerpts, and Johanning to Buz, June 30, 1904, where Johanning protested his innocence of any wrongdoing, ibid.

132. Diesel to Buz, December 20, 1905, and "Prozess gegen R. Diesel," pp. 17–30, unpublished manuscript, ibid.

133. "Verhalten des Herrn Diesel gegnüber M.A.N." This and other M.A.N. documents used by the author stem from the time of ruptured relations with Diesel and naturally exhibit a negative bias toward him.

134. Lytle, "Introduction of Diesel Power," p. 130, note 48. Lytle does not elaborate on what these plans were. The author has been unable to locate the records of the American Diesel Motor Company.

135. Johanning to Buz, February 19, 1903, Allgemeine *Nachlass*, M.A.N. Werkarchiv.

136. Letter in ibid.

137. Letter of November 14, 1902, in ibid., emphasis added.

138. Allgemeine to American Diesel Motor Company, October 23, 1902, and undated draft of a reply to the Diesel Engine Company, Ltd., both in ibid.

139. Schnauffer, "Lizenzverträge," pp. 37–38.

140. Ibid.

141. Eugen Diesel, *Diesel*, p. 389.

142. Letters in Allgemeine *Nachlass*, M.A.N. Werkarchiv.

143. Buz to Diesel, January 14, 1907, M.A.N. Werkarchiv, stated: "We will now bring a legal action against you."

144. From "Note über die schwebenden Prozess," January 2, 1909, "Prozess gegen R. Diesel, 1907–1909," and "Allgemeine Prozess gegen R. Diesel," all in M.A.N. Werkarchiv.

145. Schnauffer, "Lizenzverträge," p. 117.

146. Sass, *Geschichte*, p. 593.

147. Ibid., p. 530.

148. Eugen Diesel, *Diesel*, pp. 356–357.

149. Ibid., pp. 391–393.

150. Ibid., pp. 402–403, 417.

151. The discussion is described in Diesel, *Jahrbuch der Schiffbau-technischen Gesellschaft* (1913), pp. 355–367. Lynwood Bryant has also examined Diesel's lecture and made it the starting point of his article "Development of the Diesel Engine."

152. Diesel, *Jahrbuch* (1913), pp. 364–367.

## Conclusion

1. Bryant, "Development of the Diesel Engine," pp. 433–434, 446.

2. Linde, *Aus meinem Leben*, p. 49.

## Epilogue

1. This brief survey makes no pretense to completeness. The history of the diesel engine would require a separate volume in itself. It ought to be written some day, however, because no adequate all-inclusive account is available.

2. Gert Hack, *Alles über Diesel-Autos* (Stuttgart: Motorbuch Verlag, 1981), pp. 48–61.

3. One major area where diesels did not take hold was aviation. Numerous experiments were carried out in the United States and Europe during the interwar years, but the superior power-to-weight ratio of the gasoline engine and the advances in its technology doomed diesel aviation engines. See Constant, *Origins of the Turbojet Revolution*, pp. 134–138.

4. Gustav Goldbeck, "Geschichte des Verbrennungsmotors," Part III: "Der Dieselmotor 1900 bis 1960," *Automobil-Industrie*, 1 (1972), 37.

5. Ibid., p. 38; and Klaus Luther, "M.A.N. Diesel Engines in the History of Diesel Power Stations," 1980, unpublished manuscript in the M.A.N. Werkarchiv.

6. Sass, *Geschichte*, pp. 523–528.

7. Schnauffer, "Aus der Entwicklung des Dieselmotors," pp. 82–83.

8. Hans Flasche, "Der Dieselmaschinenbau in Russland von An-

beginn 1893 an bis etwa Jahresschluss 1918," 1947, pp. 13–14, manuscript in the Diesel *Nachlass,* Deutsches Museum.

9. Schnauffer, "Aus der Entwicklung des Dieselmotors," p. 84.

10. J. Berring, "Burmeister and Wain: Pioneers of Low Speed Marine Diesel Engines," *The Motor Ship,* April 1970, p. 10.

11. Rolf G. E. Qvarnström, "From Polar to UDAB: A Swedish Enterprise in Diesel Engines," ibid., p. 42.

12. Klaus Luther, "Erfindung, Entstehung, und Entwicklung des Dieselmotors," pp. 3–4, manuscript in the M.A.N. Werkarchiv.

13. Goldbeck, "Geschichte des Verbrennungsmotors," p. 38.

14. Ibid., pp. 39–44, 46–47; Hack, *Diesel Autos,* pp. 62–68, 72–80; Schnauffer, "Aus der Entwicklung des Dieselmotors," pp. 86–87; Jim Dunne, "Detroit's Big Switch to Turbo Power," *Popular Mechanics,* April 1984, pp. 78–81.

15. Schnauffer, "Aus der Entwicklung des Dieselmotors," p. 88; and Eugen Diesel, "Rudolf Diesel und das Automobil," essay manuscript in the Eugen Diesel *Nachlass,* Freiburg.

16. Schnauffer, "Aus der Entwicklung des Dieselmotors," p. 88.

17. *Encyclopedia Britannica,* 15th ed., s.v. "Railroads and Locomotives"; and Hans-Karl Stockklausner, *50 Jahre Diesellokomotiven* (Basel and Stuttgart: Birkhäusner, 1963), pp. 46–49.

18. Eugen Diesel, "Rudolf Diesel und das Automobil," p. 10.

19. Goldbeck, "Geschichte des Verbrennungsmotors," pp. 43–44; Hack, *Diesel Autos,* pp. 22–23; Edward Klein, "50 Jahre M.A.N. Fahrzeug Dieselmotoren," Sonderabdruck aus *Automobiltechnische Zeitschrift,* 75, Nr. 4 (April 1973), pp. 5–6.

20. The development of the diesel automobile is summarized in Hack, *Diesel Autos,* pp. 23–24, 208, 224, 226, 228, 238, 274, 316.

21. Ibid., pp. 58–60; Stephen Budiansky, "Concern Voiced over Health Effects of Diesel Exhaust," *New York Times,* January 22, 1980, C2; John Holusha, "Controlling Ignition Timing," *New York Times,* March 19, 1981, D2.

22. Hack, *Diesel Autos,* pp. 54, 57.

23. Thomas J. Lueck, "More Miles Per Gallon," *New York Times,* March 10, 1983, D2.

# Bibliographical Essay

Material for a study of Rudolf Diesel and his work is abundant. His large *Nachlass* (literary remains) is divided between the M.A.N. Werkarchiv, in Augsburg, Germany, and the Special Collections Department of the Deutsches Museum, in Munich. M.A.N.'s collection is the more extensive and includes photocopies of many of the originals in the Deutsches Museum. The M.A.N. holdings include the childhood and personal letters of Diesel to his wife and relatives as well as material relating to his social thought, to the ammonia engine, and to patents and licenses. The M.A.N. collection also contains the archives of the Dieselmotorenfabrik of Augsburg, the Allgemeine Gesellschaft für Dieselmotoren, business records of the Augsburg Engine Works, and a variety of unpublished works on the diesel engine. Eugen Diesel gave much of the personal material to M.A.N. in the late 1950s.

The Deutsches Museum possesses Diesel's theoretical manuscripts of 1892 and 1893; his correspondence with the Augsburg Engine Works, Krupp, and academic experts up to about 1896; his American diaries from 1904 and 1912; and much interesting material on the development and marketing of the engine. Diesel donated the collection to the museum shortly before his death in 1913.

Some of Diesel's personal letters from late in life are in the Eugen Diesel *Nachlass*, in the possession of Eugen's son, Rainer Diesel, in Freiburg. Because Eugen devoted considerable attention in his writings to his father's career, his *Nachlass* contains a number of essays and documents pertaining to that career. Late in life, Rudolf's daughter, Hedy, wrote her memoirs, which remain in manuscript form. The original is in the possession of her daughter, Frau Dorette Breig, Stockholm-Solluntuna, Sweden.

Diesel wrote three books and a number of essays and speeches. *Theorie und Konstruktion eines rationellen Wärmemotors* (Berlin: Julius Springer, 1893) contains most of his manuscript of 1892 plus some additional material on modifications of the ideal cycle. Diesel's and Moritz Schroeter's landmark lectures of 1897 announcing the successful tests of the diesel engine are printed in the *Zeitschrift des Vereins Deutscher*

*Ingenieure*, 41 (1897). The lectures were also published separately by the VDI in 1897 as "Diesels rationeller Wärmemotor." Diesel's *Solidarismus* (Munich: R. Oldenbourg, 1903) contains his utopian solution to workers' problems. In *Die Entstehung des Dieselmotors* (Berlin: Julius Springer, 1913), published late in his life, he explains the origins of his engine.

The chief biography of Diesel continues to be the 1937 work by his son Eugen, *Diesel: der Mensch, das Werk, das Schicksal* (Hamburg: Hanseatische Verlagsanstalt, 1937). It has been republished several times; a paperback edition, bearing the imprimatur of Heyne Verlag, Munich, appeared in 1983. Because of his personal relationship with his father and access to all his documents, Eugen was able to provide a wealth of biographical detail. He is weak, however, when examining the invention of the diesel engine and the interrelationship of his father with social and economic forces. Further, Eugen was influenced by the romantic nineteenth-century view of the inventor and presented his father as a tragic hero whose rise and fall epitomized the rise and fall of the German bourgeoisie in the late nineteenth and early twentieth centuries. Eugen's romanticized view of his father is expressed even more strongly in his *Jahrhundertwende, gesehen im Schicksal meines Vaters* (Stuttgart: Reclam, 1949). Once Eugen conceived the idea in the 1920s of writing a biography, his letters showed a more marked identification with his father (see the author's article "Diesel, Father and Son," *Technology and Culture*, 19 (July 1978), 384–385). It is not surprising, then, that Eugen's biography is less than objective.

The 1943 film, *Diesel*, on whose script Eugen worked and which was shot in Prague, is entertaining but contains as much fantasy as fact. M.A.N. was kind enough to screen it for the author in the summer of 1980.

A number of Diesel's colleagues have written memoirs or histories that deal with the diesel engine. The most significant of these is Paul Meyer's unpublished manuscript "Geschichte des Dieselmotors." He worked for Diesel's design bureau in 1898 and then transferred to the Allgemeine Gesellschaft für Dieselmotoren, Diesel's legal successor. In 1913 Meyer's *Beiträge zur Geschichte des Dieselmotors* was published by Springer, in Berlin. Diesel made a note on his copy that referred to this book as "Bureauklatsch," or office gossip, of a young engineer who did not understand the true motives of his supervisor. Later, during World War II, Meyer studied the documents available at both M.A.N. and the Deutsches Museum and wrote a much more detailed manuscript, which is one of the best studies of the diesel engine, despite its rather critical tone. Unfortunately, World War II prevented its publication. In 1945 the manuscript was rather miraculously recovered from the ruins of the VDI headquarters in Berlin. After the war, Meyer was apparently unable to interest any publishers in the manuscript.

Immanuel Lauster, an engineer who came to work for Augsburg in 1896 and subsequently was responsible for many of the improvements that turned the engine into a marketable product, also prepared a manuscript entitled "Der Dieselmotor," in the 1930s. It is quite hostile toward Diesel and must be used with caution. The head of Augsburg's steamengine division, Josef Krumper, wrote a memoir entitled "Einige Lebenserinnerungen," which contains interesting material on the period of Diesel's first approach to Augsburg in 1892. Wilhelm Worsoe's "Die Mitarbeit der Werke Fried. Krupp an der Entstehung des Dieselmotors in den Jahren 1893/97 und an den Anfangsentwicklung in den Jahren 1897/99." (Kiel: n.p., 1933, 1940) is a company history of Krupp's involvement with the diesel engine. It and an accompanying book of documents, however, are valuable for understanding the Krupp-Diesel relationship. The works of Lauster and Krumper are in the M.A.N. Werkarchiv; copies of those of Meyer and Worsoe are in both M.A.N. and the Diesel *Nachlass*, Deutsches Museum.

During the 1950s Kurt Schnauffer, now professor emeritus at the Munich Technische Universität, spent two years researching the origins of the diesel engine at the M.A.N. Werkarchiv. Frau Irmgard Denkinger, of its staff, closely collaborated in the undertaking. The result was sixteen manuscript volumes of text and documents that were part of an even larger multivolume work on the history of the internal combustion engine in Germany. Schnauffer's effort was ground-breaking, involving the most extensive examination of the archives to that time, though some of his conclusions are open to question, and his approach is narrowly technical and provides little historical framework. Even Schnauffer did not uncover all the relevant documents in the archives! The most important of his volumes for purposes of this work are: "Die Erfindung des Dieselmotors, 1890–1893," "Diesels erste Verträge, 1893," and "Lizenzverträge und Erstentwicklungen des Dieselmotors in In-und Ausland, 1893–1909."

Schnauffer's conclusions were published in "Die Erfindung Rudolf Diesels: Triumph einer Theorie," *VDI Zeitschrift*, 100 (March 1958), 308–

320; and an English synopsis in "The Invention of the Diesel Engine: the Triumph of a Theory," *Diesel-Engine-News*, No. 36, 4–22, which was also separately printed by the M.A.N. Werkarchiv. Copies of his sixteen manuscript volumes are in both the M.A.N. Werkarchiv and the Special Collections Department of the Deutsches Museum. Schnauffer's work served as the basis for Friedrich Sass's *Geschichte des deutschen Verbrennungsmotorenbaues von 1860 bis 1918* (Berlin, Göttingen, Heidelberg: Springer-Verlag, 1962). The Schnauffer and Sass works are invaluable for any study of the origins of the diesel engine.

The most significant secondary works in English are the numerous articles published by Lynwood Bryant, especially "Rudolf Diesel and his Rational Engine," *Scientific American*, 221 (August 1969), 108–117; and "The Development of the Diesel Engine," *Technology and Culture*, 17 (July 1976), 432–446. The biographies of Robert W. Nitske and Charles Morrow Wilson, *Rudolf Diesel: Pioneer of the Age of Power* (Norman, Okla.: University of Oklahoma Press, 1965), and Morton Grosser, *Diesel: The Man and the Engine* (New York: Atheneum, 1978), are in part paraphrases of Eugen Diesel's biography and are not based on original research in the German archives. C. Lyle Cummins's *Internal Fire* (Lake Oswego, Oreg.: Carnot Press, 1976) is a history of the internal combustion engine that ends with a brief chapter on Diesel. A second volume, dealing entirely with Diesel, is now being prepared. John F. Moon, *Rudolf Diesel and the Diesel Engine* (London: Priority Press, 1974), provides a brief introduction to Diesel's life and work.

The history and concerns of the German Technische Hochschulen during the nineteenth century are examined in Karl-Heinz Manegold, *Universität, Technische Hochschule, und Industrie: Ein Beitrag zur Emanzipation der Technik im 19. Jahrhundert* (Berlin: Duncker und Humblot, 1970), and in his essay "Technology Academised: Education and Training of the Engineer in the Nineteenth Century," in Wolfgang Krohn, Edwin Layton, Jr., and Peter Weingart (eds.), *Sociology of the Sciences*, Vol. II: *The Dynamics of Science and Technology* (Dordrecht, Holland: D. Reidel, 1978), pp. 137–158.

An excellent study of the German Engineers' Association and its relationship to social questions is Gerd Hortleder, *Das Gesellschaftsbild des Ingenieurs: Zum politischen Verhalten der technischen Intelligenz in Deutschland* (Frankfurt/M: Suhrkamp, 1970). A newer study of the VDI and its political-social role is Karl-Heinz Ludwig and Wolfgang König (eds.),

*Technik, Ingenieure, und Gesellschaft: Geschichte des Vereins Deutscher Ingenieure, 1856–1981* (Düsseldorf: VDI-Verlag, 1981). See also Cornelis W. R. Grispen, "Selbstverständnis und Professionalisierung deutscher Ingenieure: Eine Analyse der Nachrufe," *Technikgeschichte*, 50, Nr. 1 (1983), 34–61; and Wilhelm Treue, "Ingenieur und Erfinder: Zwei sozial- und technikgeschichtliche Probleme," *Vierteljahrsschrift für Sozial-und Wirtschaftsgeschichte*, 54 (1967), 456–476.

German engineers and their schemes to solve social problems are examined in Hans-Joachim Braun, "Ingenieure und soziale Frage, 1870–1920," *Technische Mitteilungen*, 73 (October 1980), 793–798, and (November–December 1980), 867–874. Two early works that examine the rising German engineering profession and its social role are Max Maria von Weber, "Die Stellung der Techniker im staatlichen und socialen Leben," *Wochenschrift des österreichischen Ingenieur-und Architekten-Vereins*, 2, No. 7 (February 17, 1877), 59–60, and No. 9 (March 3, 1877), 85–88; and Ludwig Brinkmann, *Der Ingenieur* (Frankfurt/M: Literarische Anstalt Rütten and Loening), 1908.

Anyone working with the Diesel papers for any length of time may rapidly despair of achieving an objective interpretation of many key events. The Diesel *Nachlass* is extensive and complex, composed of personal, technical, industrial, and business sources. Further, much has been written about Diesel that is often contradictory, hostile, or not based on original research in the archives. From the time he began to announce his new theory in 1892, he was repeatedly attacked for being unoriginal, for possessing an invalid patent, and for building an engine that did not square with his patent. By the end of his life, he was being characterized by some critics as merely a clever businessman, while M.A.N. was seen as the real inventor and developer of the engine.

Confusion existed about who originated the diesel theory, about how the engine was initially developed, and how it was later turned into a marketable product. Diesel's unwillingness to tell the full story of how his theory had been modified into a workable engine was responsible for much of the uncertainty. Undoubtedly his fear that his patents would be invalidated and that his industrial contracts would be jeopardized partly explains his reticence. Perhaps also, as Lynwood Bryant has said, Diesel was comforting himself by believing that in some way the working engine realized his original idea of achieving isothermal combustion and expansion.

Partisanship also grew rapidly among Diesel's colleagues, and their memoirs often seem self-justificatory. Diesel's own personality was unstable, and, though his personal eloquence initially won many over to his side, the long years of struggle to develop his engine and then to market it successfully caused tensions to arise in him and created a martyr complex. Eventually, he broke with many of his colleagues, including Heinrich Buz, head of M.A.N. During the first decade of the twentieth century, several lawsuits were initiated among Diesel, M.A.N., and the Allgemeine, his own creation. Last but not least, several of the key decisions in the Diesel story cannot be pinned down exactly, but can only be inferred from circumstantial evidence.

Despite some of Diesel's negative characteristics, he also exhibited traits that attract sympathy and might sway a researcher. Recognizing that some questions can never be adequately answered, this work has attempted to avoid the pitfalls of a positive or negative interpretation and to reach as far as possible to the roots of this complex story.

# Index

Printed in the United States
23867LVS00001B/65